Springer Theses

Recognizing Outstanding Ph.D. Research

Aims and Scope

The series "Springer Theses" brings together a selection of the very best Ph.D. theses from around the world and across the physical sciences. Nominated and endorsed by two recognized specialists, each published volume has been selected for its scientific excellence and the high impact of its contents for the pertinent field of research. For greater accessibility to non-specialists, the published versions include an extended introduction, as well as a foreword by the student's supervisor explaining the special relevance of the work for the field. As a whole, the series will provide a valuable resource both for newcomers to the research fields described, and for other scientists seeking detailed background information on special questions. Finally, it provides an accredited documentation of the valuable contributions made by today's younger generation of scientists.

Theses are accepted into the series by invited nomination only and must fulfill all of the following criteria

- They must be written in good English.
- The topic should fall within the confines of Chemistry, Physics, Earth Sciences, Engineering and related interdisciplinary fields such as Materials, Nanoscience, Chemical Engineering, Complex Systems and Biophysics.
- The work reported in the thesis must represent a significant scientific advance.
- If the thesis includes previously published material, permission to reproduce this must be gained from the respective copyright holder.
- They must have been examined and passed during the 12 months prior to nomination.
- Each thesis should include a foreword by the supervisor outlining the significance of its content.
- The theses should have a clearly defined structure including an introduction accessible to scientists not expert in that particular field.

More information about this series at http://www.springer.com/series/8790

Luca Brombal

X-Ray Phase-Contrast Tomography

Underlying Physics and Developments for Breast Imaging

Doctoral Thesis accepted by
University of Trieste, Trieste, Italy

 Springer

Author
Dr. Luca Brombal
Department of Physics
University of Trieste
Trieste, Italy

Supervisor
Prof. Renata Longo
Department of Physics
University of Trieste
Trieste, Italy

ISSN 2190-5053 ISSN 2190-5061 (electronic)
Springer Theses
ISBN 978-3-030-60435-6 ISBN 978-3-030-60433-2 (eBook)
https://doi.org/10.1007/978-3-030-60433-2

This Springer imprint is published by the registered company Springer Nature Switzerland AG
The registered company address is: Gewerbestrasse 11, 6330 Cham, Switzerland

Supervisor's Foreword

Modern physics contributes significantly to diagnostics and therapy in medicine: X-ray computed tomography, positron emission tomography, nuclear magnetic resonance, and particle therapy are very well-known examples. The scientific interest and the strong impact of applied physics motivate a new generation of brilliant young physicists in investigating medical applications of emerging techniques. One of them is X-ray phase-contrast imaging, which aims at exploiting the X-ray refraction by converting the phase shift into an intensity modulation to obtain an additional contrast mechanism in the recorded image. The ultimate promise of this technique is to detect low contrast details which are presently invisible to state-of-the-art clinical radiology systems. Synchrotron radiation laboratories are the headquarters of these researches, whose final goal is the translation toward compact sources.

Luca Brombal is one of the brilliant young physicists enthusiastic about medical applications; his Ph.D. research project is about X-ray phase-contrast physics and its application to improve medical imaging. This book is the scientific report of his 3 years of Ph.D. studies. The goal of the thesis is ambitious and compelling, contributing to the breast cancer diagnosis with state-of-the-art X-ray imaging techniques. The project has been developed in a European laboratory, the synchrotron radiation facility ELETTRA in Trieste (Italy), where X-ray phase-contrast imaging techniques have been developed and applied to biomedical imaging. Dr. Luca Brombal, after joining my group, quickly gained a reputation among colleagues as a brilliant physicist, equally skilled for the experimental work and the data analysis, including formal models and numerical simulations, and as a good scientific communicator. These qualities are, I believe, well represented in this Ph.D. thesis. He presents the underlying physics of X-ray phase-contrast tomography in a very plain style. However, such simplicity is the result of a deep understanding of the results published by different authors with different notations that he reported in a unified formalism. The most recent phase-contrast image formation models are explored and extended. The original results obtained during the Ph.D. project are reported in a very effective way that lets the reader understand various aspects of the interdisciplinary research. Moreover, the thesis provides the reader with practical and numerical methods to overcome the difficulties encountered in the implementation of X-ray phase-contrast imaging.

Medical Physics is often in equilibrium between basic science and public commitment, and a good project in Medical Physics should be developed in an interdisciplinary environment. This thesis well represents all these elements and I hope the reader is going to increase his/her knowledge of the exciting and challenging field of X-ray diagnostic imaging.

Trieste, Italy   Prof. Renata Longo
June 2020

Abstract

X-ray phase-contrast tomography is a powerful tool to dramatically increase the visibility of features exhibiting a faint attenuation contrast within bulk samples, as is generally the case of light (low-Z) materials. For this reason, the application to clinical tasks aiming at imaging soft tissues, as, for example, breast imaging, has always been a driving force in the development of this field. In this context, the SYRMA-3D project, which constitutes the framework of the present work, aims to develop and implement the first breast computed tomography system relying on the propagation-based phase-contrast technique at the Elettra Synchrotron facility (Trieste, Italy). This thesis finds itself in the 'last mile' towards the in-vivo application, and the obtained results add some of the missing pieces in the realization of the project, which requires multifaceted issues ranging from physical modelling to data processing and quantitative assessment of image quality to be addressed. The first part of the work introduces a homogeneous mathematical framework describing propagation-based phase contrast from the sample-induced X-ray refraction, to detection, processing and tomographic reconstruction. The original results reported in the following chapters include the implementation of a pre-processing procedure dedicated to a novel photon-counting CdTe detector; a study, supported by a rigorous theoretical model, on signal and noise dependence on physical parameters such as propagation distance and detector pixel size; hardware and software developments for improving signal-to-noise ratio and reducing the scan time; and, finally, a clinically-oriented study based on comparisons with clinical mammographic and histological images. The last part of the thesis has a wider experimental horizon, and results obtained with conventional X-ray sources are presented: a first-of-its-kind quantitative image comparison of the synchrotron-based setup against a clinically available breast-CT scanner is reported and a practical laboratory implementation of monochromatic propagation-based micro-tomography, making use of a high-power rotating anode source, is detailed.

The achieved advancements in terms of software and hardware have been significant steps towards the final goal of performing the clinical examination as effectively as possible. On the other hand, the theoretical modelling and data analysis, despite being finalized to the breast computed tomography, have a rather general validity and they can be easily extended to other propagation-based setups. The direct comparison with an existing clinical system provided further justification for the realization of the SYRMA-3D project, also suggesting the importance of synchrotron-based clinical programmes which have the potential to trigger the transition of phase-contrast imaging from synchrotrons to hospitals.

Acknowledgements

I sincerely acknowledge all the members of the SYRMA-3D collaboration and the SYRMEP group for their constant support in these fruitful years. I wholeheartedly thank my supervisor, Prof. Renata Longo, for the relentless support and trust accorded to me since the beginning of the Ph.D. I thank Prof. Luigi Rigon for having introduced me to this research field many years ago, and for the precious and insightful scientific discussions. I also thank Dr. Sandro Donato, with whom I worked countless days back-to-back acquiring, analyzing and fighting over data interpretation. Finally, I warmly thank Dr. Marco Endrizzi and Prof. Alessandro Olivo from University College London for having welcomed me into their group, giving me the possibility to understand things deeply and to work intensely from day one.

Contents

Chapter 1
Introduction

Conventional X-ray computed tomography (CT) enables the reconstruction of three-dimensional maps of X-ray attenuation properties within an investigated object, being one of the finest tools in the realm of diagnostic radiology. Anyway, when imaging low-Z samples, as soft tissues, the attenuation contrast between different materials can become faint to a point where they are no longer visible unless a large amount of radiation dose is delivered, which is unacceptable in medical diagnostic applications. This limitation has prevented a wide diffusion of breast CT imaging, where the need for high spatial and contrast resolutions required to differentiate the tissues composing the breast is hard to reconcile with a low-dose delivery, which is mandatory due to breast radiosensitivity. On the other hand, the availability of three-dimensional imaging of the breast, allowing to avoid superposition effects inherent to planar techniques (i.e. mammography), is regarded as key to improve early detection of breast cancer and/or follow-up and treatment planning stages; considering that breast cancer is one of the leading causes of death for women worldwide, this would bring to obvious clinical benefits. In this context, the use of X-ray phase-contrast imaging (XPCI) techniques can provide a major advantage over conventional attenuation-based X-ray imaging. In fact, XPCI enables to convert phase distortions (i.e. phase shift) occurring to X-ray waves travelling through a sample due to its refractive properties into detectable intensity modulations. These phase effects, which do not contribute to the image formation in conventional techniques, are in principle much stronger than attenuation, thus providing another pool of image contrast (i.e. phase contrast) and largely improving tissues visibility.

This thesis provides a detailed description of the physics underlying propagation-based phase-contrast tomography and presents several developments in terms of experimental setup, data processing and theoretical modelling towards its implementation in the field of breast imaging. The phase-contrast technique used throughout this work, namely propagation-based (PB) imaging, is arguably the simplest

© The Editor(s) (if applicable) and The Author(s), under exclusive license
to Springer Nature Switzerland AG 2020
L. Brombal, *X-Ray Phase-Contrast Tomography*, Springer Theses,
https://doi.org/10.1007/978-3-030-60433-2_1

XPCI configuration to implement experimentally, as it only requires to insert some (propagation) distance between the scanned sample and the imaging detector. On the contrary, differently from other XPCI techniques featuring more complex setups, PB imaging relies on the presence of a highly-coherent X-ray source, thus making synchrotron facilities the most suited environment for its implementation. All the experimental work has been carried out within the framework of the SYRMA-3D project, willing to perform the first synchrotron radiation-based phase-contrast breast CT at the Elettra synchrotron facility (Trieste, Italy). The main body of the thesis is organized in six chapters, whose content is summarized in the following.

• Chapter 2 is devoted at establishing the physical principles of PB imaging, from the interaction between X-ray waves and refractive objects to the phase-contrast image formation and processing, including the application of phase-retrieval algorithms and tomographic reconstruction.
• In Chap. 3 the specific challenges related to breast CT imaging are introduced and a general overview on the experimental setup is provided. In particular, many features relevant to the clinical implementation of breast CT at the SYRMEP beamline are detailed along with the specific tasks and objectives of the SYRMA-3D project.
• The main focus of Chap. 4 is the large-area CdTe photon-counting imaging detector (Pixirad-8). This detector, as many high-Z photon-counting devices, offers remarkable advantages over conventional indirect-detection charge-integration systems as high-efficiency, minimum electronic noise and spectral capabilities. Anyway, the data processing for these novel devices is still challenging mainly due to their multi-module architecture and to the presence of impurities in the sensor crystalline structure causing charge trapping. To tackle these issues an ad-hoc pre-processing software has been implemented and successfully applied to tomographic images of breast specimens.
• In Chap. 5 a theoretical model describing the effects of several physical parameters, as the propagation distance and the detector pixel size, on image noise, signal and spatial resolution is introduced and tested against experimental images. Among the results of the chapter, it is experimentally demonstrated on breast specimens that a dramatic increase in terms of signal-to-noise ratio can be achieved at a constant spatial resolution at large propagation distances, leading to the design of an extension of the beamline. At the same time, the crucial role of pixel size in determining the effectiveness of the phase retrieval, which strongly mitigates the dependence of noise on the pixel size in CT images, is quantitatively shown. Additionally, post-reconstruction phase-retrieval pipeline is introduced demonstrating that, despite the theoretical equivalence with its standard pre-reconstruction application, the proposed approach allows to eliminate artifacts in the reconstructed volume in case of acquisitions requiring multiple vertical translations.
• Chapter 6 provides a more clinically oriented focus on the imaging capabilities of the PB breast CT experimental setup. The first fully three-dimensional scans of large mastectomy samples acquired at a clinically compatible dose levels (5 mGy) and scan times (10 min) are reported and compared with conventional pla-

nar mammographic and histological images. Moreover, the possibility of further image post-processing, as 3D rendering and segmentation or bi-dimensional data compression, is investigated.

- Chapter 7 provocatively raises the question on whether it is worth to use synchrotron radiation for clinical/biomedical imaging tasks. The tentative answer is based on experimental results acquired with two setups featuring conventional rotating anode X-ray sources. In the first case, a state-of-the-art laboratory micro-CT setup, yielding monochromatic X-rays, is characterized and used in a PB configuration to image biological samples with dimensions of the order of few mm within laboratory-compatible times (from minutes to hours). This suggests that, to some extent and at a different scale, PB imaging can be implemented in a compact design even with high-power rotating anode sources. In the second case, imaging results obtained with a commercial breast CT scanner are compared with the synchrotron-based system at similar imaging conditions, showing the advantages provided by the synchrotron in terms of signal, noise, spatial resolution and, ultimately, detail visibility. Obviously these findings do not suggest that synchrotron machines should replace hospital CT scanners but, instead, that synchrotron-based studies can serve as benchmarks in terms of achievable image quality, possibly being the driving force for the development of more compact systems.

Chapter 2
Physics of Propagation-Based X-Ray Tomography

On 8 November 1895 Wilhem Conrad Röntgen discovered X-rays and, few weeks later, the famous radiograph of Mrs Röntgen's hand was imaged, marking the beginning of a new scientific discipline: radiography [1, 2]. After more than a century of unprecedented scientific, technical and technological development, clinical radiological exams, with only few exceptions, still rely on the same contrast formation mechanism, which is X-ray attenuation. Despite the immense success of conventional attenuation-based (also referred to as absorption-based) radiography and its widespread use as diagnostic tool, the advent of synchrotron radiation (SR) facilities producing intense and coherent X-ray beams allowed the researchers to focus their attention on an alternative image contrast mechanism, the phase contrast.

Phase contrast relies on the phase shift experienced by X-rays when traversing matter rather than their attenuation. In fact, the interpretation of X-rays as electromagnetic waves with a wavelength much shorter ($\sim$10,000 times) than visible light was already known at the beginning of XX century and, as stated in the far-sighted Nobel Lecture given by A. H. Compton in 1927: "[…] there is hardly a phenomenon in the realm of light whose parallel is not found in the realm of X-rays […]" [3]. This means that X-ray imaging can also take advantage of those interactions affecting the phase of the incoming wave (e.g.., refraction), which are well understood and described for visible and nearly-visible light wavelengths. The experimental arrangements allowing the detection of these effects are the so-called phase-sensitive techniques, while an image exhibiting a contrast due to phase effects is referred to as phase-contrast image.

The advent of digital detectors and powerful computers in 1970s promoted another major breakthrough in the field of diagnostic radiology, whose magnitude is comparable with the discovery of X-rays itself: computed tomography (CT) allowed for the first time to investigate bulk samples by reconstructing maps, i.e. 'slices', of their properties along the X-rays propagation plane [4]. To obtain a tomographic image, or tomogram, one needs to acquire a certain number of radiographic images,

L. Brombal, *X-Ray Phase-Contrast Tomography*, Springer Theses,
https://doi.org/10.1007/978-3-030-60433-2_2

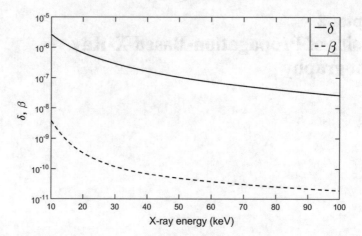

Fig. 2.1 Values of δ and β for polymethyl methacrylate (PMMA), often used as a tissue equivalent material in phantoms, between 10 and 100 keV. The semi-logarithmic plot highlights their 2–3 orders of magnitude difference spanning a broad energy range. Data from publicly available database [5]

or projections, at different angular positions of the sample. The projections are then fed into a reconstruction algorithm which inverts the tomographic problem yielding a virtually reconstructed map (or stack of maps) of the object's properties. CT was first developed in the context of conventional radiography to create X-ray attenuation maps but, given the rather general formulation of the tomographic problem, it can be in most cases straightforwardly extended to phase-contrast images, yielding, for instance, phase or even scattering maps.

This chapter is entirely devoted to explaining the physics underlying phase-contrast formation mechanism, detailing the advantages over conventional attenuation-based radiography/tomography of one of the most widespread phase-sensitive techniques, propagation-based imaging. Starting from rather general concepts, a mathematical model describing X-ray refraction will be introduced in the next section; this general model, which constitutes a common ground for many phase-sensitive techniques, will be further specialized to describe the propagation-based image formation process, also considering non perfectly coherent sources, and its inverse problem, namely the phase retrieval. Finally, the discussion will be extended to the tomographic reconstruction in the specific context of propagation-based imaging.

2.1 X-Rays Through Matter: Attenuation and Refraction

Let us consider a parallel and monochromatic beam travelling in vacuum along the z axis. In the wave formalism this can be described as a plane wave, whose space-dependent component can be written as

$$\psi = \psi_0 \, e^{ikz} \tag{2.1}$$

where ψ_0 is its real-valued amplitude, $k = |\mathbf{k}| = 2\pi/\lambda$ is the wave number and $\mathbf{k}$ is the wave vector pointing in the propagation direction, while λ is the wavelength. When the wave propagates through a medium, the wave number must be replaced by $k_{\text{medium}} = nk$, n being the complex-valued refractive index. For X-rays n is usually written as $n = 1 - \delta + i\beta$, where δ and β are real, positive and very small numbers, related, as it will be clear in the following, to the phase-shift and absorption/attenuation properties of the medium, respectively [6]. Of note, the real component of the refractive index is smaller than one, meaning that the phase-velocity in a medium is higher than the speed of light; of course this does not violate relativity as the group velocity still does not exceed the speed of light in vacuum [7]. For X-rays with energies sufficiently higher than the absorption edges of the medium, that for light materials (e.g., soft tissues) are below few keV, δ can be calculated in classical electrodynamics as

$$\delta \simeq r_0 \rho_e \lambda^2 / 2\pi \tag{2.2}$$

$r_0 = 2.82 \times 10^{-15}$ m being the classical electron radius and ρ_e the electron volume density; conversely, β is found to be proportional to λ^3 [8]. Despite being both small numbers, for biological samples and energies of interest in soft-tissue biomedical imaging (i.e. tens of keV), δ is approximately 3 orders of magnitude larger than β, their typical values being $10^{-6} - 10^{-7}$ and $10^{-9} - 10^{-10}$, respectively, as shown in Fig 2.1 [9, 10]. This huge difference is the reason why phase-sensitive techniques can be advantageous over attenuation-based imaging.

To understand how the presence of a sample can affect both amplitude and phase of the incoming X-ray wave, let us consider an object described by a three-dimensional distribution of refractive index $n(x, y, z) = 1 - \delta(x, y, z) + i\beta(x, y, z)$, traversed by the wave defined in Eq. (2.1), as schematically depicted in Fig 2.2. After the interaction with the object, the X-ray wave $\psi_{\text{out}}(x, y)$ at a given position in the object plane (x,y) will be the incident wave modulated by a complex transmission factor $T(x, y)$ [11]:

$$\psi_{\text{out}}(x, y) = \psi T(x, y) = \psi_0 e^{ikz} T(x, y) \tag{2.3}$$

where $T(x, y)$ is function of the object refractive index distribution and it is written as

$$T(x, y) = e^{ik \int (n(x,y,z)-1)\,dz} = e^{-k \int \beta(x,y,z)\,dz} \, e^{-ik \int \delta(x,y,z)\,dz} \tag{2.4}$$

with the line integral extending over the object thickness along z direction. The transmission function can be computed directly from Maxwell's equations assuming the object to be non-magnetic, with null charge and current densities [12]. Moreover, the above description implicitly assumes the so-called projection approximation to hold, meaning that the changes in the local direction of the wave vector within the sample

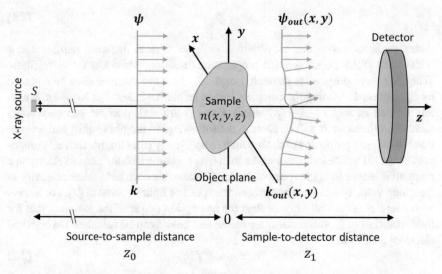

Fig. 2.2 Sketch of wave-object interaction. ψ is a monochromatic plane wave with wave vector **k** impinging on the sample described by its refractive index distribution $n(x, y, z)$. The wavefront emerging from the sample $\psi_{out}(x, y)$ is modulated both in amplitude and phase by the object and has a local wave vector $\mathbf{k}_{out}(x, y)$. z_0 and z_1 are, respectively, the source-to-sample and the sample-to-detector distances while s is the source size

are considered to be negligible. In a more pictorial description, the refraction effects are considered to be 'accumulated' through the object and to manifest themselves at its exit surface. In this way the net effect of the refractive object on the wave field can be expressed as an integral along the propagation direction of the impinging wave [13]. The previous equation implies that the object modulates the X-ray wave by reducing its amplitude by a factor dependent on β, and it introduces a shift in its phase dependent on δ, that can be written as $\Phi(x, y) = -k \int \delta(x, y, z) \, dz$.

Considering conventional radiographic techniques which are only sensitive to the transmitted X-ray intensity, i.e. the square modulus of the wave, Eq. (2.3) reduces to

$$|\psi_{out}(x, y)|^2 = |\psi_0 e^{ikz} T(x, y)|^2 = \psi_0^2 e^{-2k \int \beta(x,y,z) \, dz} \tag{2.5}$$

The latter equation can be immediately identified with the well-known Beer-Lambert law [14], describing the X-ray attenuation through an object:

$$I(x, y) = I_0 e^{-\int \mu(x,y,z) \, dz} \tag{2.6}$$

where I_0 is the beam intensity impinging on the object and $\mu = 2k\beta$ is its attenuation coefficient. At this point it is clear that in conventional imaging the phase-shift term introduced in Eq. (2.4) does not play any role at all. Conversely, the goal of any phase-sensitive technique is to detect the change in phase which, since $\delta \gg \beta$, is much bigger than attenuation.

Going back to the wave model, the phase-shift term Φ is interpreted as a local distortion of the wavefront that, at a given point of the object plane, will have a slightly different propagation direction with respect to the impinging planar wave. To determine the outgoing propagation direction at each point we assume the deviations from the initial direction z to be small (i.e. paraxial approximation) or, more formally, that the absolute values of the spatial derivatives $|(\partial/\partial x)\,\Phi(x,y)|$ and $|(\partial/\partial y)\,\Phi(x,y)|$ are much smaller than the wave number k. In this way the outgoing wave vector reads

$$\mathbf{k}_{\text{out}}(x,y) = \left(\frac{\partial}{\partial x}\Phi(x,y)\right)\hat{\mathbf{x}} + \left(\frac{\partial}{\partial y}\Phi(x,y)\right)\hat{\mathbf{y}} + k\hat{\mathbf{z}} \qquad (2.7)$$

where $\hat{\mathbf{x}}$, $\hat{\mathbf{y}}$ and $\hat{\mathbf{z}}$ are unit vectors pointing along x, y and z directions, respectively. The deviation with respect to the original direction $\hat{\mathbf{z}}$ imparted to the beam by the refractive object is expressed as a position-dependent refraction angle $\alpha(x,y)$ which is written as

$$\alpha(x,y) \simeq \frac{1}{k}\sqrt{\left(\frac{\partial}{\partial x}\Phi(x,y)\right)^2 + \left(\frac{\partial}{\partial x}\Phi(x,y)\right)^2} = \frac{1}{k}|\nabla_{xy}\Phi(x,y)| \qquad (2.8)$$

where ∇_{xy} is the gradient operator in the object plane.

Equation (2.8) is a central result of this section and provides the link between a detectable physical quantity, the refraction angle, and the object-induced phase shift. In this context, the goal of many phase-sensitive techniques will be somehow to convert this refraction angle into intensity modulations on the detector. Before describing how this can be achieved experimentally, it is worth noting that for biomedical applications (i.e. $\delta \sim 10^{-6}$ and $\lambda \sim 10^{-10}$ m) the typical refraction angles given by Eq. (2.8) range from few to few tens of microradians, hence, a posteriori, both projection and paraxial approximations hold.

2.2 The Simplest Phase-Sensitive Technique: Propagation-Based Imaging

The description of the interaction between an X-ray wave and a refractive object given so far is rather general and can serve as input to explain how many of the available phase-sensitive techniques work. As mentioned, to image the phase means to convert phase shift into intensity modulation. Broadly speaking, the plethora of techniques enabling phase imaging can be divided in into two groups, namely interferometric [15–18] and non-interferometric [19–22]. A complete description of the contrast formation mechanisms in all the phase-sensitive techniques goes beyond the scope of this work and the reader is referred to comprehensive reviews [23, 24] or books [8, 13].

In this section we focus on propagation-based (PB) imaging (note that in the literature other synonyms as in-line holography or free-space-propagation imaging can be found), which is arguably the simplest non-interferometric phase-sensitive technique to implement. Stripped down to its essence, PB imaging consists in distancing the detector from the refractive object, leaving the perturbed wavefront to propagate freely in space, as sketched in Fig. 2.2 [21]. To explain how the contrast is formed on the detector we revert our wave model to a simpler ray-tracing (or geometrical optics) approach, where X-rays are considered to be bullet-like entities whose path in each point is defined to be parallel to the local wave vector [25–27]. Moreover, it is assumed that the refractive object located in the xy plane is small compared with its distance z_1 from the image plane x_1y_1. Let be $I(x, y)$ the X-ray beam intensity emerging from the object; in the previous section we saw that this quantity is proportional to the wave square modulus, thus containing only attenuation information. Nevertheless, phase-effects manifest themselves at some propagation distance, downstream of the object. In fact, as a function of its position (x, y) on the object plane, each 'ray' is be deviated by a small angle α specified by Eq. (2.8), thus impinging on the detector at the position (x_1, y_1) given by

$$\begin{cases} x_1 \simeq x + z_1\alpha_x(x, y) \\ y_1 \simeq y + z_1\alpha_y(x, y) \end{cases} \quad (2.9)$$

where α_x and α_y are the projections of α in the planes xz and yz, respectively

$$\alpha_x = \frac{1}{k}\frac{\partial}{\partial x}\Phi(x, y) \text{ and } \alpha_y = \frac{1}{k}\frac{\partial}{\partial y}\Phi(x, y) \quad (2.10)$$

Equation (2.9) expresses simply the coordinate transformation that maps each ray from the object to the detector plane [28]. Therefore, by calculating the transformation Jacobian, one can write the intensity detected in the image plane as

$$\begin{aligned} I(x_1, y_1) &= I(x, y)\left|\frac{\partial(x_1, y_1)}{\partial(x, y)}\right|^{-1} \\ &= I(x, y)\left|\begin{matrix} 1 + z_1\frac{\partial\alpha_x}{\partial x} & z_1\frac{\partial\alpha_x}{\partial y} \\ z_1\frac{\partial\alpha_y}{\partial x} & 1 + z_1\frac{\partial\alpha_y}{\partial y} \end{matrix}\right|^{-1} \\ &\simeq I(x, y)\left(1 + \frac{z_1}{k}\nabla^2\Phi(x, y)\right)^{-1} \end{aligned} \quad (2.11)$$

where ∇^2 is the Laplacian in the object plane and the approximation is obtained by neglecting the terms $o(z_1^2\lambda^2)$. This assumption seems rather reasonable since, in a typical PB setup, z_1 is of the order of meters while $\lambda \sim 10^{-10}$ m. In those cases in which $z_1k\nabla^2\Phi(x, y) \ll 1$, i.e. when the phase contrast is 'weak' [29], a first-order Taylor expansion can be applied to Eq. (2.11), yielding

$$I(x_1, y_1) \simeq I(x, y) \left(1 - \frac{z_1}{k}\nabla^2\Phi(x, y)\right)$$
$$= I_0 e^{-2k\int \beta(x,y,z)\,dz} \left(1 - \frac{z_1}{k}\nabla^2\Phi(x, y)\right) \tag{2.12}$$

where I_0 is the X-ray intensity impinging on the object. This equation is the main result of this chapter since it explains the contrast formation principle of PB imaging. In the limit of null propagation distance $z_1 = 0$, the previous equation reduces to the Beer-Lambert law, hence only the attenuation properties of the material contribute to image formation. Conversely, by increasing z_1 another source of contrast, the phase contrast, which is proportional to the Laplacian of the phase shift, comes into play. In the case of a planar impinging wavefront, phase contrast increases linearly with the propagation distance and it is more evident at the boundaries or at sharp interfaces of the refractive object, where the phase shift changes abruptly, producing the so-called edge enhancement effect [30], as shown in Fig. 2.3. It is worth noting that, even if the ray-optical approach may be seen as a naive approximation, the same expression for intensity found in Eq. (2.12) can be demonstrated following a rigorous wave model, taking as a starting point either the (near-field) Fresnel diffraction integral or the transport-of-intensity equation [12, 13].

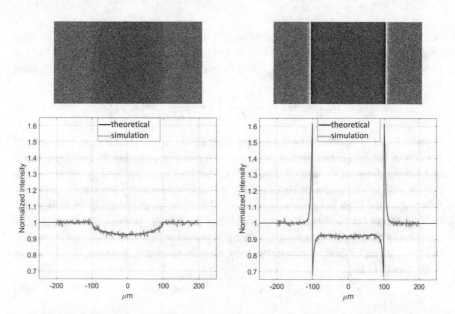

Fig. 2.3 Simulation of a 200 μm thick PMMA wire imaged at 10 keV with null propagation distance (top-left) and with 1 m of propagation distance (top-right). On the bottom the two corresponding intensity profiles matching the theoretical predictions

2.3 Effects of Finite Source Size, Detector Resolution and Near-Field Limit

So far, the whole derivation has been carried on under the hypothesis of a perfectly coherent plane wave (i.e. monochromatic and produced by a point-like source at infinite distance) and an ideal detector with a arbitrarily high spatial resolution. As it always happens, real life is sub-ideal and any deviation from both the previous assumptions can deeply affect the detected image. To study these effects let us consider a source located at a finite distance z_0 from the object plane and with a finite dimension characterized by a spatial intensity distribution PSF_{src}. At the same time let the detector be pixelated, having a finite spatial resolution and point spread function PSF_{det} which is usually of the order of one or few pixels. Let also introduce a geometrical magnification factor $M = (z_0 + z_1)/z_0$ accounting for the relative positions of source, object and detector. In this case, the detected intensity I' reads

$$I'(x_1, y_1; M) = I(x_1, y_1; M) * \left(PSF_{src} \left(\frac{x_1}{M-1}, \frac{y_1}{M-1} \right) * PSF_{det}(x_1, y_1) \right)$$

$$= I(x_1, y_1; M) * PSF_{sys}(x_1, y_1; M)$$

$$(2.13)$$

where $*$ denotes the convolution operator, PSF_{sys} is the convolution of the detector response function with the source referred to the detector plane, and $I(x_1, y_1; M)$ is the equivalent to the intensity of Eq. (2.12) when the magnification factor is accounted for [27]:

$$I(x_1, y_1; M) = \frac{I(x, y)}{M^2} \left[1 - \frac{z_1}{kM} \nabla^2 \Phi(x, y) \right] \qquad (2.14)$$

Equation (2.13) implies that the image detected in a real experiment is a blurred version of the image that would be obtained under ideal conditions and the amount of blurring depends on source distribution, detector response and geometry of the system. Given that phase-contrast manifests itself across sharp interfaces, thus contributing to the high frequency component of the image, the blurring introduced by PSF_{sys} affects primarily the phase content of the image, potentially smearing out completely the edge-enhancement effect as reported in Fig. 2.4. Taking a closer look to PSF_{sys} it can be demonstrated, by using rules of geometrical optics, that its width w goes as [31, 32]:

$$w \sim \sqrt{s^2(M-1)^2 + d^2} \qquad (2.15)$$

where s describes the source size and d the width of the detector PSF. This simple formula leads to some important considerations on the experimental implementation of PB imaging. In the majority of synchrotron-based PB experiments, the source can be considered to be ideal, meaning that its size is small and/or its distance from the object is much larger than the propagation distance (M is small): in these cases the first term in the addition of Eq. (2.15) can be neglected and the phase-contrast signal is

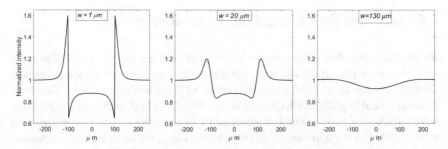

Fig. 2.4 Theoretical intensity profiles of a 200 μm thick PMMA wire convolved, from left to right, with Gaussian PSF_{sys} of full width half maximum w of 1, 20 and 130 μm respectively. To wider PSF_{sys} corresponds a loss of phase contrast due to the smearing of edge-enhancement effect

maximized by improving the detector spatial resolution and enlarging the propagation distance. On the contrary, for many conventional X-ray sources (e.g., rotating anode tubes), the source size is rather big and/or the magnification is high. In this case, any improvement in the detector resolution will not affect the visibility of phase effects since the magnitude of blurring w is dominated by the source contribution. For this reason, most of the conventional sources in use for medical applications are of no use in the field of PB phase-contrast imaging. Other practical considerations, along with the description of a dedicated PB imaging laboratory setup using a rotating anode source, can be found in Chap. 7, while more on the effects of pixel size and propagation distance is reported in Chap. 5.

Before concluding this section, some remarks on the applicability range of Eq. (2.12) should be pointed out. As stated previously, an analogous equation can be derived using the Fresnel diffraction integral in the near-field regime. This means that the given description of PB imaging technique holds for large Fresnel numbers, i.e. $N_F = a^2/(\lambda z_1) \gg 1$, where a is the smallest object's feature size of interest, which is usually related to the detector pixel size [8, 23]. This validity condition imposes an upper limit to the propagation distance (z_1) and a lower limit to the pixel size ($\sim a$), and implies that phase-contrast signal cannot be made arbitrarily large neither by increasing the propagation distance nor by decreasing the pixel size. For this reason, when setting up a PB imaging experiment, the N_F should be checked before using the aforementioned theoretical background for describing or analyzing experimental data. As an example, in the case of the experimental setup described throughout this work, a can be identified with the detector pixel size (60 μm), the propagation distance is in the order of few meters while the wavelength is a fraction of angstrom, resulting in Fresnel numbers larger than 10, so the near-field description holds. It should be noted that, conceptually, any PB imaging experimental setup can be used also in the opposite regime, i.e. far-field or Fraunhofer diffraction, provided that $N_F \ll 1$. A complete description of all the different working regimes of PB imaging can be obtained by means of the Fresnel-Kirchhoff diffraction integrals [11, 33] as illustrated in several works [16, 34, 35].

2.4   Inverting the Propagation: Phase-Retrieval

So far the image formation process in PB configuration has been described and, as a pivotal result, Eq. (2.12) was derived, expressing how the detected intensity depends on attenuation and phase properties of the illuminated object. However, many practical applications require to obtain separately both attenuation and phase-shift information rather than a phase-contrast image where their contributions are mixed [13]. The combination of this requirement with the experimentally desirable property of performing single-shot imaging results in an ill-posed problem: trying to retrieve simultaneously both phase (shift) and attenuation from Eq. (2.12) means to find solutions for two unknowns given only one equation. In the last two decades many workarounds to solve this problem, commonly known as phase-retrieval (PhR), have been derived, all of which have required multiple approximations to be made. Generally speaking, these approximations aim at reducing the number of unknowns in Eq. (2.12), thus making the expression invertible. As a first line discrimination, PhR algorithms can be split in two categories: some of them assume the sample to be non-absorbing or a 'pure phase' object, which is a suitable approximation for thin or low density samples; others require the sample to be composed of a single monomorphous material (often described as homogeneous). These and other approximations have been studied in detail in [36], listing similarities and differences between seven commonly used algorithms. In the following, a PhR algorithm falling in the second category is described and used throughout this work.

The algorithm was first proposed by Paganin and collaborators in 2002 and it is allegedly the most widely used in the PB imaging community [37]. Since this PhR technique stems from a particular version of the transport-of-intensity equation (TIE) describing a homogeneous object (TIE-Hom), it is worth starting by introducing the TIE itself [38]:

$$\nabla_{xy}\left[I(x, y; z = 0)\nabla_{xy}\Phi(x, y; z = 0)\right] = -k\frac{\partial I(x, y; z = 0)}{\partial z} \quad (2.16)$$

where, for each function of space, the z coordinate is specified to unambiguously discriminate between the object plane ($z = 0$) and the image plane ($z = z_1$). This equation provides a relation between the (measurable) intensity and the object-induced phase shift under paraxial and projection approximations. Given this definition it is not surprising that TIE is equivalent to Eq. (2.12), as demonstrated in Appendix A. The following step is to introduce the monomorphicity condition, stating that the object is composed by a single material and both δ and β (or at least their ratio) are known. In this case, phase and intensity on the object plane can be written as

$$I(x, y; z = 0) = I_0 e^{-2k\beta t(x,y)} \quad \text{and} \quad \Phi(x, y; z = 0) = -k\delta t(x, y) \quad (2.17)$$

where $t(x, y)$ is the integrated object thickness along z direction and I_0 is the X-ray intensity impinging in the object plane. The homogeneity condition allows to express both the intensity and phase terms as a function of the same variable $t(x, y)$, thus reducing the number of unknowns from two to one. Substituting the definitions of Eq. (2.17) into Eq. (2.16), and making use of the following identity

$$- k\delta\nabla_{xy}\left[e^{-2k\beta t(x,y)}\nabla_{xy}t(x,y)\right] = \frac{\delta}{2\beta}\nabla^2_{xy}e^{-2k\beta t(x,y)} \tag{2.18}$$

TIE reduces to its homogeneous version

$$\frac{\delta}{2\beta}\nabla^2_{xy}\left[I_0 e^{-2k\beta t(x,y)}\right] = -k\frac{\partial I(x,y; z=0)}{\partial z} \tag{2.19}$$

The last step of the derivation consists in finding the (approximate) expression of the derivative appearing in the right-hand side of the latter equation. Usually, it is approximated by the intensity difference between contact and image planes [12]

$$\frac{\partial I(x,y; z=0)}{\partial z} \simeq \frac{I(x,y; z=z_1) - I(x,y; z=0)}{z_1} \tag{2.20}$$

By inserting this approximation in Eq. (2.19) and re-arranging the terms we get

$$I(x,y; z=z_1) = \left(1 - \frac{z_1\delta}{2k\beta}\nabla^2_{xy}\right)I_0 e^{-2k\beta t(x,y)} \tag{2.21}$$

At this point the only unknown term is $t(x, y)$, hence TIE-Hom equation and can be solved. The solution provided by Paganin [37] makes use of the Fourier derivative theorem, yielding the projected thickness as

$$t(x, y) = -\frac{1}{2k\beta}\ln\left(\mathscr{F}^{-1}\left\{\frac{\mathscr{F}\left[I(x,y; z=z_1)/I_0\right]}{1 + \frac{z_1\delta}{2k\beta}|\mathbf{v}|^2}\right\}\right) \tag{2.22}$$

where $\mathscr{F}$ and $\mathscr{F}^{-1}$ denote the bi-dimensional Fourier transform and anti-transform, respectively, and $\mathbf{v} = (v_1, v_2)$ represents the Cartesian coordinates in the Fourier space. Once the projected thickness has been calculated it can be inserted in Eq. (2.17) to obtain both attenuation $I(x, y; z=0)$ and phase $\Phi(x, y; z=0)$ images.

The last two equations, i.e. (2.21) and (2.22), are the central result of this section; the former describes how the X-ray intensity propagate from the object to the image plane (forward propagation), the latter allows to revert this process by backpropagating (i.e. retrieving) the captured image to the object plane, as sketched in Fig. 2.5. To fully understand the effects of forward and backward propagation, it is convenient to adopt a signal processing approach where both processes are described as operators acting, respectively, on the object plane and the image plane intensity distributions [29]. From Eq. (2.21) the forward propagation operator is defined as

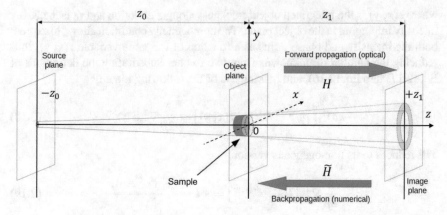

Fig. 2.5 Schematic representation of the (optical) propagation and following (numerical) phase-retrieval in a PB setup. The source plane is positioned in coordinate $-z_0$ along the z axis, the object plane defines the origin of the reference system while the image plane is positioned at z_1. H and $\tilde{H}$ denote the forward and backpropagation operators, respectively

$$H = \left(1 - \frac{z_1 \delta}{2k\beta} \nabla^2_{xy} \right) \tag{2.23}$$

which is an optical (i.e. pre-detection) deconvolution. Due the presence of the Laplace operator, H affects the image by boosting its high spatial frequency component, hence the image spatial resolution. It is worth noting that this effect, associated with PB imaging, has already been described in the previous section under the name of edge-enhancement. Conversely, the core of PhR algorithm is a bell-shaped filter in Fourier domain that, from Eq. (2.22), can be written as

$$\tilde{H} = \left(1 + \frac{z_1 \delta}{2k\beta} |\mathbf{v}|^2 \right)^{-1} \tag{2.24}$$

The effect of this filter, similar in a sense to that of a (numerical) convolution with any low-pass filter, is to reduce the image noise at cost of a worse spatial resolution [39, 40]. Anyway, the remarkable property of $\tilde{H}$ is that the resolution loss exactly compensate the spatial resolution boost due to H, i.e. to the forward propagation. Despite its apparent circularity, the combination of the forward (optical) propagation and the subsequent backward (numerical) inversion results in an image which is equivalent, up to a logarithmic transformation, to the image that would have been obtained in the object plane (i.e. the attenuation image), but with a dramatic noise reduction [41–43]. As explained by Gureyev and colleagues [29], the origin of such 'unreasonable' image quality enhancement lies in the fact that the propagation operator is an optical deconvolution (as opposed to a numerical one) which is applied prior to the image detection, thus before the generation of detection noise which is not propagated by the deconvolution itself. In terms of image quality this noteworthy effect is of paramount importance since, in general, detail visibility in any radiographic technique strongly depends on the image noise content. An experimental proof of this effect, applied to tomographic images of breast specimens, will be provided in Chap. 5.

2.5 Single- and Two-Materials Approaches to Phase Retrieval

In the derivation of the PhR filter allowing to invert TIE-Hom equation it is assumed that the investigated object is homogeneous with a known δ/β, meaning that the phase-shift and attenuation properties of the sample are proportional throughout the sample. In order to take into account the presence of two (homogeneous) materials of interest within the sample (e.g., glandular details embedded in an adipose background in breast imaging), the PhR filter reported in Eq. (2.24), referred to as single-material, can be slightly modified to

$$\tilde{H}_{2\mathrm{mat}} = \left(1 + \frac{z_1}{2k}\frac{\delta_1 - \delta_2}{\beta_1 - \beta_2}|\mathbf{v}|^2\right)^{-1} \qquad (2.25)$$

where the δ/β term has been replaced by $(\delta_1 - \delta_2)/(\beta_1 - \beta_2)$, and the subscripts refer to the two materials of interest [36]. In qualitative terms, the application of PhR allows in general to compensate for the edge-enhancement effect arising at the object interfaces upon the propagation process. Specifically, the single-material PhR allows to exactly compensate for the edge enhancement at vacuum/sample or, in practice, air/sample interfaces. Conversely, the two-materials PhR exactly compensates the edge enhancement across interfaces of two given materials embedded within the sample. In this perspective, the phase retrieval can be seen as a virtual lens which, by tuning the parameter δ/β, enables to focus upon a particular interface of interest [44]. In the case of interest of breast imaging at energies around 30 keV, δ/β is of the order of 2×10^3 for breast tissue in the single-material PhR, while $(\delta_1 - \delta_2)/(\beta_1 - \beta_2)$ is of the order of 1×10^3 for glandular/adipose interfaces in the two-materials PhR. This means that, from a signal processing perspective, the application of single-material PhR would result in a smoother image (i.e. lower noise and higher blur) with respect to the two-materials PhR [45]. Since both approaches will be used throughout this work, the type of PhR filter used will be specified for each reconstructed dataset.

2.6 Tomographic Reconstruction

While for thin bi-dimensional samples a planar image can provide sufficient information on the scanned object, for three-dimensional bulk samples (e.g., human breast), planar techniques may fail in providing an accurate description due to superposition effects. In this context, X-ray tomography is capable of overcoming such limitation, providing a fully three-dimensional map of a given object property.

A tomographic acquisition requires several planar 'views' of the sample, or projections, obtained by exposing the object to the X-ray beam at different angles. Each projection, collected at an angle θ, will be the line integral through the sample of a given object spatial distribution function $o(x, y, z)$:

$$p_\theta(x, y) = \int o(x \cos\theta - z \sin\theta, y, x \sin\theta + z \cos\theta) \, dz \qquad (2.26)$$

where the integral extends along the object thickness, y identifies the rotation axis and xz defines the tomographic plane through the object (see Fig. 2.2). Equation (2.26) identifies the Radon transform of the function $o(x, y, z)$ [46]. To reconstruct a tomographic image means to recover the spatial distribution $o(x, y, z)$ given a sufficient number of projection images $p_\theta(x, y)$ or, equivalently, to invert the Radon transform. Considering a parallel X-ray geometry, this can be accomplished by acquiring the projection images over 180 degrees and by applying the well-known filtered-back-projection (FBP) algorithm [47]:

$$o(x, y, z) = \int_0^\pi \left[\int_{-\infty}^{+\infty} P_\theta(q; y) |q| G(q) e^{2\pi i q x} \, dq \right] d\theta \qquad (2.27)$$

where $P_\theta(q; y)$ is the 1D Fourier transform of the projection p_θ along the direction x, $|q|$ is the ramp filter in the frequency domain, and $G(q)$ is the apodization filter used to limit the high spatial frequency contribution in the reconstruction. Of note, in parallel geometry, FBP does not involve the variable y, hence each reconstructed 'slice', identified by a given position y, is independent from the others.

Considering that conventional attenuation-based imaging can be seen as a special case of PB imaging at null propagation distance, rearranging Eq. (2.12) we can write

$$p_0^{\text{abs}}(x, y) = -\ln \frac{I(x, y)}{I_0} = 2k \int \beta(x, y, z) \, dz = \int \mu(x, y, z) \, dz \qquad (2.28)$$

where, for the sake of notation simplicity, the considered projection angle is $\theta = 0$. Given Eq. (2.28), the linear attenuation coefficient map $\mu(x, y, z)$ can be immediately identified with the object distribution $o(x, y, z)$ to be reconstructed by means of the FBP algorithm. The same formalism can be extended to the more general case of a finite propagation distance z_1, provided that Eq. (2.12) is conveniently re-written as

$$I(x_1, y_1) = I_0 e^{-\int \mu(x,y,z) \, dz} \left(1 - \frac{z_1}{k} \nabla_{xy}^2 \Phi(x, y) \right) \simeq I_0 e^{-\left[\int \mu(x,y,z) \, dz + \frac{z_1}{k} \nabla_{xy}^2 \Phi(x,y) \right]}$$

$$(2.29)$$

where, in the weak phase-contrast assumption, the term in parenthesis is identified with the Taylor expansion of an exponential term [48]. Starting from the previous expression, and recalling that $\Phi(x, y) = -k \int \delta(x, y, z) \, dz$, the projection image acquired in PB configuration reads

$$p_0^{PB}(x, y) = -\ln \frac{I(x_1, y_1)}{I_0} = \int \mu(x, y, z) \, dz - z_1 \nabla_{xy}^2 \int \delta(x, y, z) \, dz \quad (2.30)$$

In this case, the tomographic reconstruction of the first term provides the attenuation coefficient map whereas the second term corresponds to the three-dimensional Laplacian of the decrement from unity of the refractive index $\delta(x, y, z)$. In summary, for PB imaging, the reconstructed distribution is approximated by

$$o^{PB}(x, y, z) = \mu(x, y, z) - z_1 \nabla_{xyz}^2 \delta(x, y, z) \quad (2.31)$$

Equation (2.31) is of great importance since it proves that, similarly to the planar case, a tomographic map reconstructed from PB projections will be similar to the (conventional) attenuation map except for object interfaces or sharp edges, where the (three-dimensional) Laplacian of δ is expected to be significantly different from zero.

Finally, the tomographic reconstruction of phase-retrieved projections should be considered. Following the Paganin's approach, in the derivation of the PhR formula the imaged object is assumed to be homogeneous, so its attenuation and phase-shift properties (or at least their ratio) are constant throughout the volume. The application of the phase retrieval yields, for each projection, the object projected thickness, which, given the homogeneity assumption, is proportional to the line integrals of both $\mu(x, y, z)$ and $\delta(x, y, z)$.

$$p_0^{PhR}(x, y) = t(x, y) = \frac{1}{\mu_{in}} \int \mu(x, y, z) \, dz = \frac{1}{\delta_{in}} \int \delta(x, y, z) \, dz \quad (2.32)$$

where the proportionality constants $1/\mu_{in}$ and $1/\delta_{in}$ are input parameters of the PhR filter as reported in Eq. (2.22). Given this definition of the projection image, the tomographic reconstructed quantity will be

$$o^{PhR}(x, y, z) = \frac{1}{\mu_{in}} \mu(x, y, z) = \frac{1}{\delta_{in}} \delta(x, y, z) \quad (2.33)$$

Of note, starting from phase-retrieved projections, the reconstructed image is found to be proportional to the (conventional) attenuation image $\mu(x, y, z)$, meaning that the image contrast is equal to the attenuation contrast. In case of medical applications, this is of great importance since tomographic images reconstructed after applying the PhR procedure can be calibrated in terms of linear attenuation coefficients, which is the standard procedure in conventional X-ray tomography [49]. More details on the phase-retrieval effects on the reconstructed image will be discussed in Chap. 5.

References

1. Brailsford JF (1946) Roentgen's discovery of x rays: their application to medicine and surgery. Br J Radiol 19(227):453–461. https://doi.org/10.1259/0007-1285-19-227-453
2. Mould RF (1980) A History of X-rays and radium. IPC Building & ContraCT Journals Limited
3. Compton AH (1927) Nobel prize-winner tells of discoveries: X-rays as a branch of optics. Sci News-Lett 12(349):387–388. https://doi.org/10.2307/3902440
4. Hounsfield GN (1980) Computed medical imaging. J Comput Assis Tomogr 4(5):665. https://doi.org/10.1097/00004728-198010000-00017 (Nobel lecture, 8 Dec 1979)
5. Taylor JA (2018) TS imaging. http://ts-imaging.science.unimelb.edu.au/Services/Simple/
6. Als-Nielsen J, McMorrow D (2011) Elements of modern X-ray physics. Wiley, New York. https://doi.org/10.1002/9781119998365
7. Griffiths DJ (2017) Introduction to electrodynamics.
8. Rigon L (2014) X-ray imaging with coherent sources. In: Brahme A (ed) Comprehensive biomedical physics, vol 2. Elsevier, pp 193–216. https://doi.org/10.1016/B978-0-444-53632-7.00209-4
9. Lewis R (2004) Medical phase contrast X-ray imaging: current status and future prospects. Phys Med Biol 49(16):3573. https://doi.org/10.1088/0031-9155/49/16/005
10. Zhou S-A, Brahme A (2008) Development of phase-contrast X-ray imaging techniques and potential medical applications. Phys Med 24(3):129–148. https://doi.org/10.1016/j.ejmp.2008.05.006
11. Born M, Wolf E (1999) Principles of optics, 7th (expanded) edition. Cambridge University Press, Cambridge, UK, p 890. https://doi.org/10.1017/CBO9781139644181
12. Paganin D (2006) Coherent X-ray optics. Number 6 in Oxford series on synchrotron radiation. Oxford University Press on Demand. https://doi.org/10.1093/acprof:oso/9780198567288.001.0001
13. Pelliccia D, Kitchen MJ, Morgan KS (2018) Theory of X-ray phase-contrast imaging. In: Russo P (ed) Handbook of X-ray imaging: physics and technology. Taylor and Francis, pp 971–997. ISBN 978-1-4987-4152-1. https://doi.org/10.1201/9781351228251 (chapter 49)
14. Cunningham JR, Johns HE (1983) The physics of radiology. Springfield: Charles C. Thosmas. https://doi.org/10.1118/1.595545
15. Bonse U, Hart M (1965) An X-ray interferometer. Appl Phys Lett 6(8):155–156. https://doi.org/10.1063/1.1754212
16. Snigirev A, Snigireva I, Kohn V, Kuznetsov S, Schelokov I (1995) On the possibilities of X-ray phase contrast microimaging by coherent high-energy synchrotron radiation. Rev Sci Instrum 66(12):5486–5492. https://doi.org/10.1063/1.1146073
17. Momose A (1995) Demonstration of phase-contrast X-ray computed tomography using an X-ray interferometer. Nucl Instrum Methods Phys Res Sect A: Accelerators, Spectrometers, Detectors Assoc Equipment 352(3):622–628. https://doi.org/10.1016/0168-9002(95)90017-9
18. Cloetens P, Guigay J, De Martino C, Baruchel J, Schlenker M (1997a) Fractional talbot imaging of phase gratings with hard x rays. Opt Lett 22(14):1059–1061. https://doi.org/10.1364/OL.22.001059
19. Davis T, Gao D, Gureyev T, Stevenson A, Wilkins S (1995) Phase-contrast imaging of weakly absorbing materials using hard x-rays. Nature 373(6515):595. https://doi.org/10.1038/373595a0
20. Ingal V, Beliaevskaya E (1995) X-ray plane-wave topography observation of the phase contrast from a non-crystalline object. J Phys D Appl Phys 28(11):2314. https://doi.org/10.1088/0022-3727/28/11/012
21. Wilkins S, Gureyev TE, Gao D, Pogany A, Stevenson A (1996) Phase-contrast imaging using polychromatic hard x-rays. Nature 384(6607):335. https://doi.org/10.1038/384335a0
22. Olivo A, Arfelli F, Cantatore G, Longo R, Menk R, Pani S, Prest M, Rigon L, Tromba G, Vallazza E et al (2001) An innovative digital imaging set-up allowing a low-dose approach to phase contrast applications in the medical field. Med Phys 28(8):1610–1619. https://doi.org/10.1118/1.1388219

23. Bravin A, Coan P, Suortti P (2012) X-ray phase-contrast imaging: from pre-clinical applications towards clinics. Phys Med Biol 58(1):R1. https://doi.org/10.1088/0031-9155/58/1/R1

24. Olivo A, Castelli E (2014) X-ray phase contrast imaging: from synchrotrons to conventional sources. Rivista del nuovo cimento 37(9):467–508. https://doi.org/10.1393/ncr/i2014-10104-8

25. Ishisaka A, Ohara H, Honda C (2000) A new method of analyzing edge effect in phase contrast imaging with incoherent x-rays. Opt Rev 7(6):566–572. https://doi.org/10.1007/s10043-000-0566-z

26. Monnin P, Bulling S, Hoszowska J, Valley J-F, Meuli R, Verdun F (2004) Quantitative characterization of edge enhancement in phase contrast X-ray imaging. Med Phys 31(6):1372–1383. https://doi.org/10.1118/1.1755568

27. Peterzol A, Olivo A, Rigon L, Pani S, Dreossi D (2005) The effects of the imaging system on the validity limits of the ray-optical approach to phase contrast imaging. Med Phys 32(12):3617–3627. https://doi.org/10.1118/1.2126207

28. Gureyev T, Wilkins S (1998) On X-ray phase imaging with a point source. JOSA A 15(3):579–585. https://doi.org/10.1364/JOSAA.15.000579

29. Gureyev TE, Nesterets YI, Kozlov A, Paganin DM, Quiney HM (2017) On the "unreasonable" effectiveness of transport of intensity imaging and optical deconvolution. JOSA A 34(12):2251–2260. https://doi.org/10.1364/JOSAA.34.002251

30. Spanne P, Raven C, Snigireva I, Snigirev A (1999) In-line holography and phase-contrast microtomography with high energy x-rays. Phys Med Biol 44(3):741. https://doi.org/10.1088/0031-9155/44/3/016

31. Gureyev TE, Nesterets YI, Stevenson AW, Miller PR, Pogany A, Wilkins SW (2008) Some simple rules for contrast, signal-to-noise and resolution in in-line X-ray phase-contrast imaging. Opt Express 16(5):3223–3241. https://doi.org/10.1364/OE.16.003223

32. Brombal L, Kallon G, Jiang J, Savvidis S, De Coppi P, Urbani L, Forty E, Chambers R, Longo R, Olivo A et al (2019b) Monochromatic propagation-based phase-contrast microscale computed-tomography system with a rotating-anode source. Phys Rev Appl 11(3):034004. https://doi.org/10.1103/PhysRevApplied.11.034004

33. Cowley JM (1995) Diffraction physics. Elsevier. https://doi.org/10.1016/B978-0-444-82218-5.X5000-7

34. Pogany A, Gao D, Wilkins S (1997) Contrast and resolution in imaging with a microfocus X-ray source. Rev Sci Instrum 68(7):2774–2782. https://doi.org/10.1063/1.1148194

35. Arfelli F, Assante M, Bonvicini V, Bravin A, Cantatore G, Castelli E, Dalla Palma L, Di Michiel, M, Longo R, Olivo A et al (1998) Low-dose phase contrast X-ray medical imaging. Phys Med Biol 43(10):2845. https://doi.org/10.1088/0031-9155/43/10/013

36. Burvall A, Lundström U, Takman PA, Larsson DH, Hertz HM (2011) Phase retrieval in X-ray phase-contrast imaging suitable for tomography. Opt Express 19(11):10359–10376. https://doi.org/10.1364/OE.19.010359

37. Paganin D, Mayo S, Gureyev TE, Miller PR, Wilkins SW (2002) Simultaneous phase and amplitude extraction from a single defocused image of a homogeneous object. J Microsc 206(1):33–40. https://doi.org/10.1046/j.1365-2818.2002.01010.x

38. Teague MR (1983) Deterministic phase retrieval: a green's function solution. JOSA 73(11):1434–1441. https://doi.org/10.1364/JOSA.73.001434

39. Barrett HH, Myers KJ (2003) Foundations of image science. Wiley, New York. https://doi.org/10.1117/1.1905634

40. Gureyev T, Nesterets Y, de Hoog F (2016) Spatial resolution, signal-to-noise and information capacity of linear imaging systems. Opt Express 24(15):17168–17182. https://doi.org/10.1364/OE.24.017168

41. Nesterets YI, Gureyev TE (2014) Noise propagation in X-ray phase-contrast imaging and computed tomography. J Phys D Appl Phys 47(10):105402. https://doi.org/10.1088/0022-3727/47/10/105402

42. Kitchen MJ, Buckley GA, Gureyev TE, Wallace MJ, Andres-Thio N, Uesugi K, Yagi N, Hooper SB (2017) CT dose reduction factors in the thousands using X-ray phase contrast. Sci Rep 7(1):15953. https://doi.org/10.1038/s41598-017-16264-x

43. Brombal L, Donato S, Dreossi D, Arfelli F, Bonazza D, Contillo A, Delogu P, Di Trapani, V, Golosio B, Mettivier G et al (2018b) Phase-contrast breast CT: the effect of propagation distance. Phys Med Biol 63(24):24NT03. https://doi.org/10.1088/1361-6560/aaf2e1

44. Beltran M, Paganin D, Uesugi K, Kitchen M (2010) 2D and 3D X-ray phase retrieval of multi-material objects using a single defocus distance. Opt Express 18(7):6423–6436. https://doi.org/10.1364/OE.18.006423

45. Brombal L, Golosio B, Arfelli F, Bonazza D, Contillo A, Delogu P, Donato S, Mettivier G, Oliva P, Rigon L et al (2018c) Monochromatic breast computed tomography with synchrotron radiation: phase-contrast and phase-retrieved image comparison and full-volume reconstruction. J Med Imaging 6(3):031402. https://doi.org/10.1117/1.JMI.6.3.031402

46. Deans SR (2007) The Radon transform and some of its applications. Courier Corporation. https://doi.org/10.1016/S0079-6638(08)70123-9

47. Buzug TM (2011) Computed tomography. In: Springer handbook of medical technology. Springer, Berlin, pp 311–342. https://doi.org/10.1007/978-3-540-39408-2

48. Cloetens P, Pateyron-Salomé M, Buffiere J, Peix G, Baruchel J, Peyrin F, Schlenker M (1997b) Observation of microstructure and damage in materials by phase sensitive radiography and tomography. J Appl Phys 81(9):5878–5886. https://doi.org/10.1063/1.364374

49. Piai A, Contillo A, Arfelli F, Bonazza D, Brombal L, Cova MA, Delogu P, Trapani VD, Donato S, Golosio B, Mettivier G, Oliva P, Rigon L, Taibi A, Tonutti M, Tromba G, Zanconati F, Longo R (2019) Quantitative characterization of breast tissues with dedicated CT imaging. Phys Med Biol 64(15):155011. https://doi.org/10.1088/1361-6560/ab2c29

Chapter 3
Propagation-Based Breast CT and SYRMA-3D Project

So far, the physical foundations of propagation based imaging have been introduced. Making use of those concepts, we now steer our attention on the specific application of PB technique in the field of breast computed tomography (BCT). As aforementioned, phase-contrast imaging is appealing for discriminating soft tissues featuring a poor attenuation contrast. This means that, by using conventional X-ray techniques, soft details embedded in a different soft tissue background are, in general, hard to detect. This is the case of breast cancer diagnosis, where one aims at detecting glandular or tumoral details embedded in an adipose background.

Breast cancer is one of the most frequently diagnosed malignancies and one of the leading causes of death for women worldwide [1, 2]. Early detection is therefore a key factor for adequately treating and defeating this disease, leading to more successful treatment and, ultimately, to a decrease of the associated mortality and morbidity. For these reasons, application of phase-contrast imaging to breast cancer detection has always been a driving force in the development of phase-sensitive techniques, potentially leading to benefits of major clinical relevance. In the last two decades dozens of studies are reported on this issue and more are yet to come. Anyway, mainly due to the stringent requirements on the source coherence and/or constraints on the delivered radiation dose, only two clinical mammographic (i.e. planar imaging) studies using phase-contrast techniques have been performed so far. Both of them take advantage of the PB configuration and, while the first is based on conventional X-ray tubes [3], the second makes use of synchrotron radiation (SR) [4]. Despite starting from the same physical principles, the outcomes of the trials largely differ. In fact, in the first case, limitations inherent to the conventional X-ray system were found to overwhelm the advantages related to phase contrast and, after a clinical trial encompassing 3835 examinations, no statistically significant difference was found in recall rates and cancer detection rates when compared to conventional film-screening mammography [5]. On the contrary, the SR based mammography trial, encompassing more than 70 patients, demonstrated better image quality [6], lower delivered dose and higher diagnostic power with respect to digital mammography [7]. In addition

L. Brombal, *X-Ray Phase-Contrast Tomography*, Springer Theses, https://doi.org/10.1007/978-3-030-60433-2_3

to phase-contrast effects, breast cancer diagnosis can take advantage of tomographic systems aiming at overcoming the tissue superposition inherent in planar techniques, potentially hindering the detection of massive lesions.

In this chapter some concepts of breast computed tomography will be introduced and the SYRMA-3D project, which constitutes the framework of this thesis, will be described.

3.1 Breast CT

At present, the most widely used clinical tool for the early diagnosis of breast cancer is 2D digital mammography (DM). As aforementioned, in addition to the low attenuation contrast between soft tissues composing the breast, mammography is also affected by tissue superposition. With the aim of reducing the masking effect of tissue overlap, a pseudo three-dimensional (sometimes referred to as 2.5D) imaging technique, namely digital breast tomosynthesis (DBT) [8], has been developed in the last decade, showing to offer some diagnostic advantages over DM even for screening purposes [9–11]. In any case, for both DM and DBT, the breast has to be compressed to reduce its thickness, thus limiting tissue superposition, often resulting in a severe discomfort for the patient. In fact, up to 76% of women experience pain or discomfort during a mammographic procedure and moderate level of pain can persist up to four days post-examination, sometimes even discouraging from the participation to the screening program [18]. On the contrary, BCT is, in principle, capable of providing a fully three-dimensional map of the X-ray attenuation properties of the non-compressed organ, thus entirely avoiding anatomical noise and patient discomfort [12–17].

To image a non-compressed breast means to increase the X-ray energy in order to have a sufficient transmission through the organ. Increasing the energy from values suitable for mammography (around 20 keV) to values adequate for tomography (around 30 keV) determines a further contrast reduction in the attenuation properties of soft tissues, thus requiring high contrast sensitivity to provide tissue differentiation. Of note is that, at mammographic energies, the attenuation coefficient of soft tissues is dominated by the photoelectric contribution, which is very sensitive to small atomic number (Z) differences (proportional to Z^4), while at tomographic energies it is dominated by the Compton cross section, which has a shallower dependence on the atomic number (proportional to Z) [19]. More details on trade-offs between contrast and X-ray transmission in BCT will be given in Chap. 6 and they can be found in [20]. In addition to image contrast, one of the arguably biggest challenges for BCT is to match high spatial resolution with a low dose CT exam: the optimization of these two conflicting requirements is one of the major issues that the medical community is facing nowadays (see also Sect. 5.2). Specifically, high spatial resolution is mainly required to detect the presence of microcalcifications, which are tiny deposits of calcium potentially being early signs of malignancy; at the same time, the need for a low radiation dose examination is dictated by the high radiosensitivity of breast tissue.

The above, among other technical difficulties, had held up the development of BCT with respect to general purpose body-CT scanners. In fact, even if the first clinical studies in the field of BCT were published 10 years ago [13], the technique is not yet established in the radiological community. At present, only two BCT scanners are available on the market [15, 21–24], but their use is not widespread and their role in the diagnostic process has still to be fully recognized [25–27]. The first generation of BCT scanners is based on cone beam geometry [28] which, while keeping the acquisition time quite short, suffers a contrast reduction due to scattered radiation [29]. In order to overcome such limitation a new generation of BCT systems, based on fan beam and photon-counting detectors, has been recently developed [17, 22, 30]. The fan beam setup adds complexity to the system requiring a spiral-CT acquisition and potentially longer scan times, suggesting the usefulness of a breast immobilizer devices [31].

In this context, the synchrotron-based experimental setup used described in the next sections adopts a configuration conceptually more similar to the new generation BCT systems, encompassing a laminar beam and a photon-counting detector.

3.2 The SYRMEP Beamline

All the experimental breast-imaging activities presented in this thesis have been carried out at the SYnchrotron Radiation for MEdical Physics (SYRMEP) beamline at the Italian synchrotron facility Elettra (Trieste, Italy). Elettra is a third generation synchrotron where electrons of energy of either 2.0 or 2.4 GeV circulate in a 260 m long storage ring. Making use of various sources, such as bending magnets, ondulators and wigglers, synchrotron radiation is extracted from the accelerated electrons, feeding 26 experimental beamlines tangentially positioned with respect to the storage ring. Each beamline is dedicated to a different X-ray technique, offering the users a huge variety of options to probe their samples such as spectroscopy, spectromicroscopy, diffraction, scattering and lithography.

At SYRMEP a widespread research activity in bio-medical imaging has been developed since 1997 [32–34]; besides conventional attenuation-based imaging, several phase-contrast techniques such as PB and analyzer-based imaging have been successfully applied and developed. In this regards, one of the core programs of the beamline is the development of clinical studies in the field of breast imaging making use of a PB experimental setup. As mentioned in the previous section, the world's first SR mammographic study has been completed in 2011 at SYRMEP, showing that SR PB mammographic images yield better diagnostic performances with respect to conventional imaging [4]. Those encouraging results led soon thereafter to a new project aiming at upgrading the existing setup, allowing BCT to be implemented [30, 35, 36].

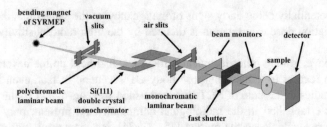

Fig. 3.1 A schematic overview of the SYRMEP beamline, from the X-ray production by the bending magnet (left) to the sample and detector stages (right). Some additional elements present in the beam's path, e.g., anti-scattering grids, are not reported in figure

A schematic overview of the beamline is shown in Fig. 3.1. The light source is a bending magnet providing a polychromatic (white) beam highly collimated (laminar) in the vertical direction (divergence of the order of 10^{-4} rad). The white beam goes through a couple of orthogonal tungsten vacuum slits, defining the horizontal acceptance of 7 mrad. The beam is then optionally monochromatized by means of a double Si(1,1,1) crystal working in Bragg configuration, allowing to tune the beam energy in the range 8–40 keV, with an effective energy resolution of $\Delta E/E = 2 \times 10^{-3}$. Thanks to a recent upgrade, the monochromator insertion has been automatized so that the user can switch from white to monochromatic beam configurations in few minutes. The monochromatic beam footprint is further adjusted by another set of tungsten slits before traversing a pair of beam monitors (air ionization chambers) which are read simultaneously. In case of any misbehaviour of the beam or discrepancy between the readings of the two ionization chambers, the beam is promptly stopped by a fast shutter system, operating in 15 ms [37]. Finally, the beam reaches the sample and, after a propagation distance of 1.6 m (in the present configuration), it reaches the detector. This arrangement results in a laminar beam with a useful cross section of about 220 (horizontal)$\times$3.5 mm^2 (vertical, Gaussian shape, FWHM) at 32 keV and source distance of 30 m, where the patient support is located. More on the concept of useful cross section and on the energy dependence of the vertical dimension of the beam will be discussed in Sect. 5.3. The available monochromatic flux largely depends on the selected energy and on the synchrotron operation mode, i.e. 2.0 or 2.4 GeV, as shown in Fig. 3.2. Taking as a reference energy 17 keV, the maximum flux at 2.0 GeV with a typical ring current of 300 mA is 1.5×10^8 photons/mm^2/s, while at 2.4 GeV and current of 180 mA, it is more than 4 times higher, namely 7×10^8 photons/mm^2/s. For energies around 30 keV, which is of interest for the BCT application, the only feasible operating condition is 2.4 GeV since the available flux at 2.0 GeV is more than one order of magnitude lower. In addition, it should be remarked that the X-ray source is extremely stable thanks to the top-up operating mode of the synchrotron, meaning that the ring current is kept constant (fluctuations generally well below 1 %) through frequent electron injections compensating for the natural ring current decay. Of course this property is of great importance in sight of clinical applications.

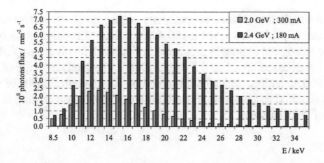

Fig. 3.2 Monochromatic flux as a function of the selected energy at 30 m from the source available at the SYRMEP beamline. Different colors identify different operation modes of the synchrotron, i.e. electrons stored at 2.0 GeV or 2.4 GeV

3.3 The SYRMA-3D Project

SYnchrotron Radiation for MAmmography (SYRMA-3D) is a project founded by the Italian National Institute of Nuclear Physics (INFN) in collaboration with Elettra. The project aims at achieving the first clinical application of propagation-based phase-contrast breast computed tomography (PBBCT), making use of the PB imaging setup available at the SYRMEP beamline [30, 35, 36, 38]. The activity of the SYRMA-3D collaboration includes all the topics necessary for the implementation of a clinical study, ranging from an *ad-hoc* Monte Carlo simulation software for dose evaluation to the development of a dedicated image quality assessment procedure. Both the experimental setup and data processing entail a number of novelties, each of which requires a dedicated study and optimization. In the following sections, a general overview of the experiment, from sample stage to image reconstruction, is provided. It is worth mentioning here that SYRMA-3D is not the only synchrotron radiation-based BCT program: a project with similar ambitions and methodologies is presently undergoing at the Imaging and Medical Beamline (IMBL) at the Australian Synchrotron (Melbourne, Australia) and there is a longstanding collaboration between Italian and Australian research teams [39].

3.4 Patient Support

After being monochromatized and filtered, the laminar X-ray beam enters a dedicated experimental room, defined as patient room. In the patient room a specifically developed support is positioned in the beam propagation direction. It is made by a rotating support with an ergonomically designed aperture at the rotation center where the samples or, in the future, the patient's breast, can be imaged in a pendant geometry as shown in the left panel of Fig. 3.3 [28]. The patient support ensures a constant rotation speed, which is fundamental for CT acquisitions, and it allows horizontal and vertical translations of several centimeters with a precision better than 100 μm. A single scan is performed when a 180 deg continuous rotation is completed: for most of the tomographic images presented in this work this is accomplished in 40 s by setting the rotation speed to 4.5 deg/s. Due to the laminar shape of the beam, a multi-scan acquisition, typically composed of 10–15 vertical steps of the patient support, is required to scan a significant portion of the breast, leading to an overall scan time of 7–10 minutes.

3.5   Imaging Configuration and Detector

Images are collected at the largest propagation distance presently available, 1.6 m, which is sufficient to detect phase-contrast effects and, along with the laminar shape of the beam, to avoid scattering contribution not requiring anti-scattering grids or dedicated scattering-removal algorithms [35]. The imaging device is a novel large-area high-efficiency photon-counting detector featuring a global active area of 246×25 mm^2 which fits well the beam profile and it is shown in the right panel of Fig. 3.3 [40]. A thorough description of the detector will be given in Chap. 4. For the sake keeping the scan duration as short as possible, the detector is always operated at the maximum available frame rate of 30 Hz, corresponding to 1200 evenly-spaced projections over the 40 s long rotation. In this context, it is worth mentioning that the detector frame rate is the main bottleneck in speeding up the acquisition. In fact, both the beam flux and the patient support would be able to cope with a two-fold higher rotation speed, which would still result in an acceptable patient comfort.

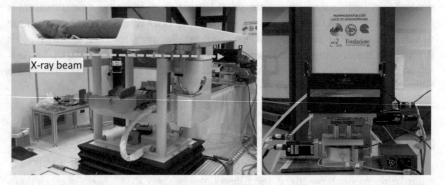

Fig. 3.3 Pictures of the experimental setup in the patient room: the rotating support holding an anthropomorphic phantom hanging from the ergonomically designed hole at the center of rotation (left) and a frontal view of the Pixirad-8 detector (right)

3.6 Dose Control System and Dosimetric Quantities

When dealing with patients being exposed to X-rays, the dosimetric control system is of paramount importance to ensure image acquisitions at acceptable radiation dose levels. To this end, a custom dosimetric system, previously developed for the mammography program, is used in the PBBCT project. It is based on two custom-made high-precision ionization chambers (see Fig. 3.1) positioned approximately 3 m upstream from the sample (breast). The chambers measure the entrance radiation dose in terms of absolute air kerma, that is used to define exposure parameters. Within a wide energy range (9–40 keV), the ionization chambers are calibrated against the standard air kerma chamber for low-energy X-rays by the Department of Ionizing Radiation Metrology of the Italian National Agency for New Technologies, Energy and Environment (ENEA) [41–43]. In a clinical scenario, the dosimetric system allows to measure the entrance radiation dose in terms of absolute air kerma in real time throughout the examination. In case of any accidental event, potentially altering the predetermined level of radiation dose or compromising the image quality, the safety system is designed to promptly interrupt the examination by triggering a fast shutter (described previously), thus ensuring the patient's safety [37]. The radiation flux can be finely tuned by filtering the beam with electro-actuated aluminum sheets, allowing the insertion of filtration thicknesses ranging from 0.125 to 7.875 mm. More on a novel filtration system available at the SYRMEP beamline will be detailed in Sect. 5.3.

The exposure parameters of the irradiated object, i.e. X-ray flux and energy, are usually chosen in order to match a given dosimetric quantity, that is, in our case, the total mean glandular dose (MGD_t). MGD_t is defined as the ratio between the total energy deposited in the whole breast and the glandular mass in the irradiated volume, as opposed to mean glandular dose (MGD), that is the mean dose to the glandular mass present in the whole breast [44]. It should be noted that, when irradiating only a portion of the breast, MGD_t is a conservative dosimetric quantity since it accounts also for the energy scattered outside the irradiated volume. Moreover, as reported in Fig. 3.4, MGD_t varies very slowly changing the thickness of irradiated region and it converges to MGD when the entire breast is irradiated. Thus, MGD_t is an appropriate dosimetric quantity in a clinical scenario where only a partial scan of the breast is required (e.g., due to acquisition time constraints or previous knowledge of a specific region of interest). In practice, MGD_t is calculated by multiplying the air kerma at breast position by a conversion factor accounting for breast size and glandularity, obtained from an *ad-hoc* developed Monte Carlo simulation based on a GEANT4 code optimized for breast dosimetry [44, 45].

So far, the dose reference value for the clinical exam in the SYRMA-3D project has been $MGD_t = 5$ mGy, which is lower than (or comparable to) the existing BCT systems but slightly higher than a standard two-view mammography. Anyway, expected image quality improvements due to an imminent upgrade of the beam-line (see Chap. 5) would allegedly lower the reference value to 2 mGy, which is comparable to standard mammography.

Fig. 3.4 MGD (square) and MGD$_t$ (triangle) as a function of the height of the irradiated volume, resulting from a Monte Carlo simulation of a cylindrical breast phantom (50% glandular fraction) with a diameter of 10 cm and a height of 7.5 cm, and a beam energy of 32 keV. MGD converges to MGD$_t$ upon irradiation of the whole phantom [44]

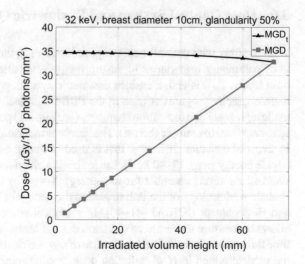

3.7   Data Processing and Image Quality Control Assessment

Once the projection images are collected by the detector, they are streamed to a dedicated safe storage and adequately cropped to save disk space: each scan requires approximately 1 Gb of memory. Data then undergo an *ad-hoc* pre-processing procedure, described in Chap. 4, aiming at compensating detector related artifacts. At this point the corrected projections are loaded onto a custom reconstruction software where the phase-retrieval filter is applied and a suitable GPU-based tomographic reconstruction algorithm (both standard filter back projection or iterative procedures) can be selected, yielding the final reconstructed image displayed as a stack of slices [46].

As required to any clinical imaging system, the SYRMA-3D experimental setup needs a quality control protocol to ensure high image quality and homogeneous results among the examinations. Since the imaging system differs from any clinical system, a dedicate image quality assessment procedure has been developed, making use of a custom phantom composed by several rods of different tissue-like materials and filled with water [47]. The phantom allows to perform absolute image calibration in terms of attenuation coefficients, to evaluate accuracy and reproducibility, to test image uniformity, noise fluctuations and low contrast resolution [48].

Once tiled together, all the aforementioned elements form a complete exam workflow, that is depicted schematically in Fig. 3.5.

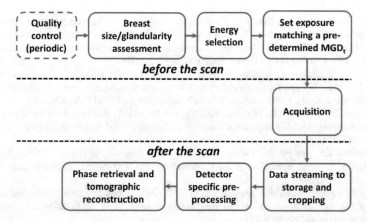

Fig. 3.5 Scheme of the exam workflow from the preliminary periodic quality control to the tomographic reconstruction

References

1. Ferlay J, Soerjomataram I, Dikshit R, Eser S, Mathers C, Rebelo M, Parkin DM, Forman D, Bray F (2015) Cancer incidence and mortality worldwide: sources, methods and major patterns in GLOBOCAN 2012. Int J Cancer 136(5):E359–E386. https://doi.org/10.1002/ijc.29210
2. Siegel RL, Miller KD, Jemal A (2016) Cancer statistics, 2016. CA Cancer J Clin 66(1):7–30. https://doi.org/10.3322/caac.21332
3. Tanaka T, Honda C, Matsuo S, Noma K, Oohara H, Nitta N, Ota S, Tsuchiya K, Sakashita Y, Yamada A et al (2005) The first trial of phase contrast imaging for digital full-field mammography using a practical molybdenum X-ray tube. Invest Radiol 40(7):385–396. https://doi.org/10.1097/01.rli.0000165575.43381.48
4. Castelli E, Tonutti M, Arfelli F, Longo R, Quaia E, Rigon L, Sanabor D, Zanconati F, Dreossi D, Abrami A et al (2011) Mammography with synchrotron radiation: first clinical experience with phase-detection technique. Radiology 259(3):684–694. https://doi.org/10.1148/radiol.11100745
5. Morita T, Yamada M, Kano A, Nagatsuka S, Honda C, Endo T (2008) A comparison between film-screen mammography and full-field digital mammography utilizing phase contrast technology in breast cancer screening programs. In: International workshop on digital mammography. Springer, pp 48–54. https://doi.org/10.1007/978-3-540-70538-3_7
6. Longo R, Tonutti M, Rigon L, Arfelli F, Dreossi D, Quai E, Zanconati F, Castelli E, Tromba G, Cova MA (2014) Clinical study in phase-contrast mammography: image-quality analysis. Phil Trans R Soc A 372(2010):20130025. https://doi.org/10.1098/rsta.2013.0025
7. Fedon C, Rigon L, Arfelli F, Dreossi D, Quai E, Tonutti M, Tromba G, Cova MA, Longo R (2018) Dose and diagnostic performance comparison between phase-contrast mammography with synchrotron radiation and digital mammography: a clinical study report. J Med Imaging 5(1):013503. https://doi.org/10.1117/1.JMI.5.1.013503
8. Sechopoulos I (2013) A review of breast tomosynthesis. Part I, the image acquisition process. Med Phys 40(1). https://doi.org/10.1118/1.4770279
9. McDonald ES, Oustimov A, Weinstein SP, Synnestvedt MB, Schnall M, Conant EF (2016) Effectiveness of digital breast tomosynthesis compared with digital mammography: outcomes analysis from 3 years of breast cancer screening. JAMA Oncol 2(6):737–743. https://doi.org/10.1001/jamaoncol.2015.5536

10. Lång K, Andersson I, Rosso A, Tingberg A, Timberg P, Zackrisson S (2016) Performance of one-view breast tomosynthesis as a stand-alone breast cancer screening modality: results from the malmö breast tomosynthesis screening trial, a population-based study. Eur Radiol 26(1):184–190. https://doi.org/10.1007/s00330-015-3803-3

11. Phi X-A, Tagliafico A, Houssami N, Greuter MJ, de Bock GH (2018) Digital breast tomosynthesis for breast cancer screening and diagnosis in women with dense breasts—a systematic review and meta-analysis. BMC Cancer 18(1):380. https://doi.org/10.1186/s12885-018-4263-3

12. Boone JM, Nelson TR, Lindfors KK, Seibert JA (2001) Dedicated breast CT: radiation dose and image quality evaluation. Radiology 221(3):657–667. https://doi.org/10.1148/radiol.2213010334

13. Lindfors KK, Boone JM, Nelson TR, Yang K, Kwan AL, Miller DF (2008) Dedicated breast CT: initial clinical experience. Radiology 246(3):725–733. https://doi.org/10.1148/radiol.2463070410

14. Lindfors KK, Boone JM, Newell MS, D'Orsi CJ (2010) Dedicated breast computed tomography: the optimal cross-sectional imaging solution? Radiol Clin 48(5):1043–1054. https://doi.org/10.1016/j.rcl.2010.06.001

15. O'Connell A, Conover DL, Zhang Y, Seifert P, Logan-Young W, Lin C-FL, Sahler L, Ning R (2010) Cone-beam CT for breast imaging: Radiation dose, breast coverage, and image quality. Am J Roentgenol 195(2):496–509. https://doi.org/10.2214/AJR.08.1017

16. Sechopoulos I, Feng SSJ, D'Orsi CJ (2010) Dosimetric characterization of a dedicated breast computed tomography clinical prototype. Med Phys 37(8):4110–4120. https://doi.org/10.1118/1.3457331

17. Kalender WA, Beister M, Boone JM, Kolditz D, Vollmar SV, Weigel MC (2012) High-resolution spiral CT of the breast at very low dose: concept and feasibility considerations. Eur Radiol 22(1):1–8. https://doi.org/10.1007/s00330-011-2169-4

18. Papas MA, Klassen AC (2005) Pain and discomfort associated with mammography among urban low-income african–american women. J Commun Health 30(4):253–267. https://doi.org/10.1007/s10900-005-3704-5

19. Evans RD (1955) The atomic nucleus. McGraw-Hill New York. https://doi.org/10.1002/aic.690020327

20. Delogu P, Di Trapani V, Brombal L, Mettivier G, Taibi A, Oliva P (2019) Optimization of the energy for breast monochromatic absorption X-ray computed tomography. Sci Rep 9(1):1–10. https://doi.org/10.1038/s41598-019-49351-2

21. Koning C (2018) Koning breast CT. http://koninghealth.com/en/kbct/

22. Kalender WA, Kolditz D, Steiding C, Ruth V, Lück F, Rößler A-C, Wenkel E (2017) Technical feasibility proof for high-resolution low-dose photon-counting CT of the breast. Eur Radiol 27(3):1081–1086. https://doi.org/10.1007/s00330-016-4459-3

23. Berger N, Marcon M, Saltybaeva N, Kalender WA, Alkadhi H, Frauenfelder T, Boss A (2019) Dedicated breast computed tomography with a photon-counting detector: initial results of clinical in vivo imaging. Invest Radiol. https://doi.org/10.1097/RLI.0000000000000552

24. AB-CT (2019) Advanced breast-CT. https://www.ab-ct.com/nuview/

25. O'Connell AM, Karellas A, Vedantham S (2014) The potential role of dedicated 3d breast CT as a diagnostic tool: review and early clinical examples. Breast J 20(6):592–605. https://doi.org/10.1111/tbj.12327

26. Wienbeck S, Lotz J, Fischer U (2017) Review of clinical studies and first clinical experiences with a commercially available cone-beam breast CT in Europe. Clin Imaging 42:50–59. https://doi.org/10.1016/j.clinimag.2016.11.011

27. Uhlig J, Uhlig A, Biggemann L, Fischer U, Lotz J, Wienbeck S (2019) Diagnostic accuracy of cone-beam breast computed tomography: a systematic review and diagnostic meta-analysis. Eur Radiol 29(3):1194–1202. https://doi.org/10.1007/s00330-018-5711-9

28. Sarno A, Mettivier G, Russo P (2015) Dedicated breast computed tomography: basic aspects. Med Phys 42(6Part1):2786–2804. https://doi.org/10.1118/1.4919441

29. Sechopoulos I (2012) X-ray scatter correction method for dedicated breast computed tomography. Med Phys 39(5):2896–2903. https://doi.org/10.1118/1.4711749

30. Longo R, Arfelli F, Bellazzini R, Bottigli U, Brez A, Brun F, Brunetti A, Delogu P, Di Lillo F, Dreossi D et al (2016) Towards breast tomography with synchrotron radiation at elettra: first images. Phys Med Biol 61(4):1634. https://doi.org/10.1088/0031-9155/61/4/1634
31. Rößler A, Wenkel E, Althoff F, Kalender W (2015) The influence of patient positioning in breast CT on breast tissue coverage and patient comfort. Senologie-Zeitschrift für Mammadiagnostik und-therapie 12(02):96–103. https://doi.org/10.1055/s-0034-1385208
32. Arfelli F, Assante M, Bonvicini V, Bravin A, Cantatore G, Castelli E, Dalla Palma L, Di Michiel M, Longo R, Olivo A et al (1998) Low-dose phase contrast X-ray medical imaging. Phys Med Biol 43(10):2845. https://doi.org/10.1088/0031-9155/43/10/013
33. Abrami A, Arfelli F, Barroso R, Bergamaschi A, Bille F, Bregant P, Brizzi F, Casarin K, Castelli E, Chenda V et al (2005) Medical applications of synchrotron radiation at the syrmep beamline of elettra. Nucl Instrum Methods Phys Res Sect A Accel Spectrom Detectors and Assoc Equip 548(1–2):221–227. https://doi.org/10.1016/j.nima.2005.03.093
34. Tromba G, Longo R, Abrami A, Arfelli F, Astolfo A, Bregant P, Brun F, Casarin K, Chenda V, Dreossi D (2010). The syrmep beamline of elettra: clinical mammography and bio-medical applications. In: AIP conference proceedings, vol 1266. AIP, pp 18–23. https://doi.org/10.1063/1.3478190
35. Brombal L, Golosio B, Arfelli F, Bonazza D, Contillo A, Delogu P, Donato S, Mettivier G, Oliva P, Rigon L et al (2018c) Monochromatic breast computed tomography with synchrotron radiation: phase-contrast and phase-retrieved image comparison and full-volume reconstruction. J Med Imaging 6(3):031402. https://doi.org/10.1117/1.JMI.6.3.031402
36. Longo R, Arfell F, Bonazza D, Bottigli U, Brombal L, Contillo A, Cova M, Delogu P, Di Lillo F, Di Trapani V et al (2019) Advancements towards the implementation of clinical phase-contrast breast computed tomography at elettra. J Synchrotron Radiat 26(4). https://doi.org/10.1107/S1600577519005502
37. Longo R, Abrami A, Arfelli F, Bregant P, Chenda V, Cova MA, Dreossi D, De Guarrini F, Menk R, Quai E et al (2007) Phase contrast mammography with synchrotron radiation: physical aspects of the clinical trial. In: International Society for Optics and Photonics on Medical Imaging 2007: Physics of Medical Imaging, vol 6510, p 65100T. https://doi.org/10.1117/12.708403
38. Delogu P, Golosio B, Fedon C, Arfelli F, Bellazzini R, Brez A, Brun F, Di Lillo F, Dreossi D, Mettivier G et al (2017b) Imaging study of a phase-sensitive breast-CT system in continuous acquisition mode. J Instrum 12(01):C01016. https://doi.org/10.1088/1748-0221/12/01/C01016
39. Taba ST, Baran P, Lewis S, Heard R, Pacile S, Nesterets YI, Mayo SC, Dullin C, Dreossi D, Arfelli F et al (2019) Toward improving breast cancer imaging: radiological assessment of propagation-based phase-contrast CT technology. Acad Radiol 26(6):e79–e89. https://doi.org/10.1016/j.acra.2018.07.008
40. Bellazzini R, Spandre G, Brez A, Minuti M, Pinchera M, Mozzo P (2013) Chromatic X-ray imaging with a fine pitch cdte sensor coupled to a large area photon counting pixel asic. J Instrum 8(02):C02028. https://doi.org/10.1088/1748-0221/8/02/C02028
41. Burns D, Toni M, Bovi M (1999) Comparison of the air-kerma standards of the enea-inmri and the bipm in the low-energy X-ray range. Technical Report BIPM-99/11, Bureau International des Poids et Mesures
42. Bovi M, Laitano R, Pimpinella M, Toni M, Casarin K, Quail E, Tromba G, Vacotto A, Dreossi D (2007) Absolute air-kerma measurement in a synchrotron light beam by ionization free-air chamber. In: Workshop on Absorbed dose and air kerma primary standards, Paris
43. Bovi M, Laitano R, Pimpinella M, Toni M, Casarin K, Tromba G, Vascotto A (2009) Misure assolute di kerma in aria del fascio di luce di sincrotrone prodotto presso l'impianto elettra di trieste per applicazioni nella diagnostica medica ad alta risoluzione. In: Proceedings of the 6th Conference "Metrologia & Qualità", Turin, Italy, pp 7–9
44. Mettivier G, Fedon C, Di Lillo F, Longo R, Sarno A, Tromba G, Russo P (2015) Glandular dose in breast computed tomography with synchrotron radiation. Phys Med Biol 61(2):569. https://doi.org/10.1088/0031-9155/61/2/569

45. Fedon C, Longo F, Mettivier G, Longo R (2015) Geant4 for breast dosimetry: parameters optimization study. Phys Med Biol 60(16):N311. https://doi.org/10.1088/0031-9155/60/16/N311
46. Brun F, Pacilè S, Accardo A, Kourousias G, Dreossi D, Mancini L, Tromba G, Pugliese R (2015) Enhanced and flexible software tools for X-ray computed tomography at the italian synchrotron radiation facility elettra. Fundamenta Informaticae 141(2–3):233–243. https://doi.org/10.3233/FI-2015-1273
47. Contillo A, Veronese A, Brombal L, Donato S, Rigon L, Taibi A, Tromba G, Longo R, Arfelli F (2018) A proposal for a quality control protocol in breast CT with synchrotron radiation. Radiol Oncol 52(3):1–8. https://doi.org/10.2478/raon-2018-0015
48. Piai A, Contillo A, Arfelli F, Bonazza D, Brombal L, Cova MA, Delogu P, Trapani VD, Donato S, Golosio B, Mettivier G, Oliva P, Rigon L, Taibi A, Tonutti M, Tromba G, Zanconati F, Longo R (2019) Quantitative characterization of breast tissues with dedicated CT imaging. Phys Med Biol 64(15):155011. https://doi.org/10.1088/1361-6560/ab2c29

Chapter 4
Detector and Pre-processing

As mentioned in the previous chapter, the SYRMA-3D experimental setup encompasses a novel CdTe photon-counting detector. This imaging tool can offer great advantages over conventional X-ray detectors but, at the same time, requires careful characterization and specific processing to attain high-quality artifact-free images. In this context, the main goals of the present chapter are to provide the detector characterization and to demonstrate the effectiveness of the implemented pre-processing procedure. Many of the contents presented in the following have been published in [1].

4.1 Photon-Counting Detectors: An Overview

In recent years high-Z large-area single-photon-counting detectors have become appealing for imaging applications both in synchrotron and conventional sources experiments [2]. These detectors offer remarkable advantages over conventional indirect detection and charge integration systems. Properly operated high-Z single-photon-counting detectors show minimum electronic noise (i.e. noise is Poisson dominated), energy discrimination of photons (i.e. spectral performances) and high detective efficiency [3, 4]. Moreover, unlike scintillator-based detectors where an increase in the efficiency typically leads to a decrease in the spatial resolution due to the scintillating process regardless of the pixel dimension, in direct conversion devices the spatial resolution is mainly limited by the pixel size [5]. The aforementioned features make these detectors suitable for low dose phase-contrast imaging experiments, where both high efficiency for limiting the dose and high spatial resolution to detect phase effects (e.g., edge enhancement) are needed.

L. Brombal, *X-Ray Phase-Contrast Tomography*, Springer Theses,
https://doi.org/10.1007/978-3-030-60433-2_4

At present, however, the data processing of large area high-Z single-photon-counting detectors is still challenging. In fact, given the limited area of a single sensor (typically few cm^2) a large field of view is obtained via a multi-module architecture employing arrays or matrices of sensors [6, 7]. These arrangements lead to the presence of non-negligible gaps between the sensors and, when the sample footprint is bigger than a single module, to the use of close-to-edge pixels which often show worse efficiency, stability and gain constancy. Moreover, when dealing with modular detectors, both the alignment of the sensors, possibly leading to image distortions, as well as the energy threshold equalization among the modules can be critical. In addition, these detectors usually suffer from local charge-trapping effects due to impurities in the sensor's crystalline structure. Charge trapping is in general, dependent on the polarization time and on the exposure [8–11], leading to severe ring artifacts in CT applications, where the scan duration may be in the order of several seconds or more [12]. In absence of a dedicated pre-processing procedure, all these effects cause artifacts which alter significantly the image quality, possibly impairing its scientific or diagnostic significance.

4.2   Pixirad-8

The imaging device used in the SYRMA-3D experimental setup is a large-area CdTe photon-counting detector (Pixirad-8), produced by Pixirad s.r.l., an INFN spin-off company [1, 13, 14]. The basic building block of the detector features a pixelated solid-state CdTe sensor which is connected to a matching CMOS readout ASIC via the flip-chip bonding technique. Pixirad-8 is made up by an array of 8 modules tiled together, each one with an active area of 30.7×24.8 mm^2, leading to a global active area of 246×24.8 mm^2. The pixels are arranged on a honeycomb matrix with 60 μm pitch, corresponding to a pixel-to-pixel spacing of 60 μm in the horizontal direction and 52 μm in the vertical direction, leading to an overall matrix dimension of 4096×476 pixels [12]. Each pixel is associated with two independent discriminators and 15-bit counters which can be used either in color or in dead-time-free mode. The first mode, suitable for polychromatic X-ray spectra applications, allows to set two different energy thresholds, thus enabling spectral imaging [15]. Conversely, when the second mode is selected, which is always the case throughout this work, both discriminators are set to the same threshold and one counter is filled while the other is being read, thus providing a virtually dead-time-free acquisition. This modality allows to perform continuous acquisitions where the organ is constantly irradiated without delivering unnecessary radiation dose and not needing any beam-shutter/detector synchronization.

Considering a beam energy of 30 keV and a detector threshold of 5 keV, resembling the working condition for the images presented in this work, the detector is linear up to approximately 2×10^5 counts/pixel/s, corresponding to 6.4×10^7 counts/mm^2/s, as shown in the left panel of Fig. 4.1. Moreover, given the CdTe sensor thickness of 650 μm, the detector has an absorption efficiency higher than 99.9% up to 40 keV.

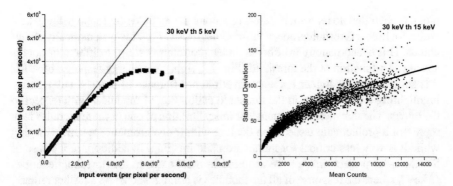

Fig. 4.1 Pixirad linearity with a beam energy of 30 keV and a threshold of 5 keV (left panel): points are the experimental data and solid line represents an ideal linear response. Standard deviation as a function of the mean counts in 50 pixels regions with a beam energy of 30 keV and threshold of 15 keV (right panel): scattered points are experimental data and line is the ideal Poissonian noise. Reproduced from [14] by permission of IOP Publishing

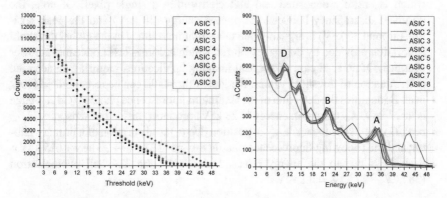

Fig. 4.2 Module-by-module threshold scan (left panel) and corresponding differential spectra (right panel), obtained with a beam energy of 38 keV. The origin of 4 peaks in the differential spectra, labelled with letters *A-D*, is explained in text

When the detector threshold is set to half of the beam energy in order to limit multiple counts due to charge-sharing effects, the detector noise is found to follow the Poisson statistics, i.e. it is equal to the square root of the counts, as reported in the right panel of Fig. 4.1.

Pixirad-8 allows to define only a global energy threshold, so it sets the same threshold for all the modules. Anyway, when a multi-module architecture is involved, differences among the detector blocks may arise, leading to discrepancies in the energy response larger than the intrinsic energy resolution of each sensor. With the aim of testing the threshold homogeneity, a threshold scan at a fixed beam energy of 38 keV has been carried out and the mean count values of each module have been plotted as a function of the global energy threshold (Fig. 4.2, left panel). From the plot it is clear that no relevant differences among the modules are found below 9 keV while, for higher thresholds, one of the modules (ASIC 7) yields systematically higher counts. This phenomenon can be better understood considering the differential counts spectra as a function of the threshold (Fig. 4.2, right); the energy response of ASIC 7 is shifted towards higher energies (maximum difference of about 8 keV), hence highlighting a discrepancy in the threshold calibration of module 7 with respect to the others. The threshold discrepancy or miscalibration of one or more modules may represent a problematic issue when dealing with (polychromatic) spectral imaging while it is way less critical for monochromatic imaging. In addition, in the case of SYRMA-3D project, images are acquired in most cases at low threshold energies (3 keV), where the response of all the modules is homogeneous. As a further remark, it should be noted that each of the differential spectra features 4 peaks (from A to D in figure) which demonstrate the energy resolving capabilities of the detector and provide an interesting insight into the involved physical processes:

- peak A, also defined as full-energy peak, represents the impinging photon energy which is entirely deposited and collected within a single pixel. Of note, the observed discrepancy between the photon energy (38 keV) and the peak position (around 35 keV) is arguably due to a slight inaccuracy in the global threshold calibration provided by the manufacturer, which becomes less noticeable at lower energies. Of note, in case of spectral imaging applications requiring an accurate determination of threshold values, the global threshold can be re-calibrated [15].
- peak B identifies the detection of the Cadmium fluorescence photons (Cd $K-$edge is at 27 keV, K_α transition energy is 23 keV) produced inside the sensor;
- peak C reflects the local energy deposition due to the absorption of a primary photon and the following fluorescence photon escape (38 keV $-$ 23 keV $=$ 15 keV);
- peak D is due to the energy released locally by the $K-$shell photoelectron (38 keV $-$ 27 keV $=$ 11 keV).

4.3 Pre-processing Procedure: Description

The SYRMA-3D collaboration put a great effort in the realization of a multi-step pre-processing procedure dedicated to the Pixirad-8 detector to obtain 'clinical-like' images not containing potential confounding factors due to the presence of artifacts. As a general remark, it is worth noting that the relevance and the interplay among various sources of artifacts are dependent on the specific application. For instance, time-dependent effects as charge trapping may be of little or no importance for fast planar imaging, while being detrimental in CT; on the contrary, the effect of insensitive gaps between detector modules can be easily compensated for in CT, where lost information is recovered at different projection angles, while it can be critical in planar imaging. For this reason, albeit being specific for the SYRMA-3D experiment, the implemented pre-processing procedure has a modular structure allowing to adapt or modify each module independently to cope with specific experimental requirements. In the following, a detailed description of this procedure is given and the effects of each step both on projections and reconstructed images are documented.

The term pre-processing refers to all the elaborations on raw data needed, regardless of the acquisition parameters, to compensate for detector-related artifacts, yielding, in this case, a set of corrected projections ready to be phase-retrieved and reconstructed. The modular structure of the pre-processing procedure comprises five steps, namely dynamic flat fielding, gap seaming, dynamic ring removal, projection despeckling and around-gap equalization. Each of these steps require as input several parameters that have been optimized on actual breast specimens datasets, in order to mimic a realistic clinical scenario.

For the sake of computational efficiency and portability, the code is implemented in C language. The complete processing of a typical experimental dataset comprising 1200 16-bit raw projections, with a dimension of 2300×70 pixels each, requires about 4 minutes on a 8 cores Intel Core i7-6700 CPU @ 3.40 GHz including loading and saving operations.

4.3.1 Dynamic Flat Fielding

The flat-fielding procedure is common to most of the X-ray imaging applications and it serves multiple purposes, namely to correct beam shape and intensity inhomogeneities, to equalize different gain levels among pixels and to perform an image normalization, preparing planar projections for CT reconstruction. The standard flat fielding consists of a pixel-by-pixel division of each projection image with a constant flat image (i.e. acquired without the sample). Defining $P(x, y; t)$ as the projection image, with x, y the pixel coordinates and t the projection index proportional to the acquisition time, and $\bar{F}_0(x, y)$ the constant flat image, the corrected image with a standard procedure will be

$$f_{\text{static}}(x, y; t) = \frac{P(x, y; t)}{\bar{F}_0(x, y)} \tag{4.1}$$

Given a fixed detector frame rate, the statistical noise of $\bar{F}_0(x, y)$ is decreased by computing the average of $(2w + 1)$ flat images, where w determines the width of the averaging window

$$\bar{F}_0(x, y) = \frac{1}{2w + 1} \sum_{t=1}^{2w+1} F(x, y; t) \tag{4.2}$$

The choice of an odd number as the window width has been made for the sake of notation coherence: in the following most of the presented filter windows are centered in a pixel of interest so that an odd filter dimension is required. With this procedure, hereinafter referred to as static flat fielding, the presence of a detector gain time dependence in the projection images cannot be compensated since the flat image is not time dependent. On the contrary, the implemented dynamic flat-fielding approach requires as many flat-field images as the number of projections so that the denominator of Eq. (4.1) can be substituted with a moving average of $2w + 1$ flat images

$$\bar{F}(x, y; t) = \frac{1}{2w + 1} \sum_{t'=t-w}^{t+w} F(x, y; t') \tag{4.3}$$

In this way, if the gain time dependence is reproducible, each flat image has both a high statistics and the same time dependence as the projection images. The flat fielded projections will be

$$f_{\text{dynamic}}(x, y; t) = \frac{P(x, y; t)}{\bar{F}(x, y; t)} \tag{4.4}$$

In order for this approach to be used, a slow time dependence of gain is assumed so that, within the moving average window $2w + 1$, the flat images are considered to be constant. Namely, given a 30 Hz frame rate and a window $2w + 1 = 11$ frames, the gain should not vary significantly for times in the order of 1 s. In addition, the fluctuations of the beam are assumed to be small in the time scale of the acquisition: this requirement is generally fulfilled at the Elettra synchrotron operated in top-up mode, where 1 mA of ring current is injected every 20 minutes, having a baseline of 140 to 180 mA at 2.4 GeV.

As a further remark, it is worth mentioning that a different approach to dynamic flat fielding exists, and it is based on principal component analysis [16]. This technique, often used to compensate for instabilities due to vibrations or drifts in the beam, generally requires a smaller number of flat-field images to capture intensity variations, being advantageous when the scan time is long and/or the number of projections is high (see Sect. 7.2.2). Anyway, in the specific case discussed in this chapter, the scan time is short and the acquisition of as many flat-field images as the number of projections would add only 40 s to the whole examination workflow. Moreover, as it will be clear in the next section, the detector gain variations are relatively smooth and the acquisition of many flat-field images has proven to be insightful in understanding the time-dependence of the mentioned detector gain drifts.

4.3.2 Gap Seaming

As most of multi-module single-photon-counting devices, the Pixirad-8 detector has a small gap (3 pixels wide) between adjacent modules which needs to be filled within the pre-processing procedure. The selected approach is a linear interpolation with a rectangular 9×8 pixels kernel. For each pixel within the gap, the interpolation window is chosen to be half in the left module and half in the right one (regions A and B in Fig. 4.3), then the mean value of each half is computed and the gap-pixel

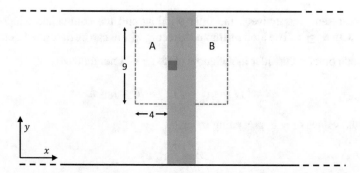

Fig. 4.3 Illustration of the gaps seaming procedure. The gray region represents the gap while the rectangle is the interpolation window used for the pixel of interest (dashed line). The figure is not to scale

value is defined as the weighted average of the two mean values

$$f_{\text{gap}}(x, y; t) = \frac{u(x)}{N_A} \sum_{(x',y')\in A} f(x', y'; t) + \frac{v(x)}{N_B} \sum_{(x',y')\in B} f(x', y'; t) \qquad (4.5)$$

where $N_A = N_B$ is the normalization factor while the weights $u(x)$ and $v(x)$ are the normalized distances between the pixel within the gap and the regions A and B. Despite its simplicity, this procedure represents a good compromise between image quality and computational load. Nevertheless, more sophisticated approaches, such as the inpainting technique described in [17], may be considered if wider gaps or high-contrast sample details crossing two modules are present.

4.3.3 Dynamic Ring Removal

Ring artifacts are produced by gain inhomogeneities at the pixel level and they are commonly encountered in tomographic reconstruction. In most of the cases the pixel (or group of pixels) producing the ring has a constant gain offset with respect to its neighbors, so that a single equalization is sufficient to remove or at least mitigate the artifact. In this case, despite the application of the dynamic flat fielding, some pixels still show a time dependent gain, resulting in rings with a non-constant intensity. To compensate for these artifacts a dynamic (i.e. depending on the projection index) equalization factor has to be used. The implemented ring-removal algorithm makes use of the alpha-trimmed filter, which is a hybrid of the mean and median filters [18]. For each pixel, this filter takes a window of nearest neighbors, sorts their values, excludes the largest and the smallest values and replaces the pixel with the average of the remaining ones. Let $g(i)$ be a one-dimensional image, h and c two integers which represent, respectively, the filter window and the confidence window half widths, with $h \geq c$. The alpha-trimmed filter algorithm can be described as follows:

- For each pixel i, consider the window of its $2h + 1$ neighbors

$$w(j) = g(i + j), \quad -h \leq j \leq h \tag{4.6}$$

- Sort the values of w in ascending order

$$w_s = \text{sort}(w) \tag{4.7}$$

- Substitute the pixel i with the average of w_s within the confidence window of size $2c + 1$

$$\bar{g}_s(i) = \frac{1}{2c + 1} \sum_{j=-c}^{c} w_s(j) \tag{4.8}$$

Basically, in this average we are excluding the $h - c$ smallest values and the $h - c$ largest values. Note that if $c = 0$ the alpha-trimmed filter reduces to the median filter, while if $c = h$ it reduces to the mean filter. In a two-dimensional or three-dimensional image, the alpha-trimmed filter can be applied along each dimension: we will call $S_x[g]$, $S_y[g]$ and $S_t[g]$ the images filtered along the dimensions x, y and t respectively. Furthermore, we define the filter applied along two or three dimensions as the composition of two or three one-dimensional alpha-trimmed filters, as for instance $S_{xy}[g]=S_x[S_y[g]]$.

Given $f(x, y; t)$, the three-dimensional matrix describing the whole set of projections, and $G_t^\sigma[f]$, the convolution of the image f with a Gaussian function of standard deviation σ along the projection axis t, the ring removal algorithm consists of the following steps:

- First apply the alpha-trimmed filter to the projections along the dimension t, then filter them with a Gaussian convolution along the same dimension

$$f_1(x, y; t) = G_t^\sigma[S_t[f]](x, y; t) \tag{4.9}$$

where σ should be a significant fraction of the number of projections.
- Apply the alpha-trimmed filter to f_1 along the dimensions x and y

$$f_2(x, y; t) = S_{xy}[f_1](x, y; t) \tag{4.10}$$

- f_1 is smooth along the dimension t by construction. It is also expected to be a smooth function along the dimensions x and y, therefore f_2 and f_1 should be close to each other, unless there is an equalization problem. Evaluate the equalization correction factor as

$$\alpha(x, y; t) = f_2(x, y; z)/f_1(x, y; t) \tag{4.11}$$

- Apply the correction factor to obtain the ring-corrected image

$$f_{\text{rc}}(x, y; t) = \alpha(x, y; t) f(x, y; t) \qquad (4.12)$$

In our implementation, we are using $2h + 1 = 10$, $2c + 1 = 5$ for all dimensions and $\sigma = N_p/10$. Here we remark that the main advantage of this algorithm is that the equalization factor α varies with the projection index, allowing to cope with non-constant ring artifacts. The results of this approach will be compared with two of the most known filters which tackle the ring-removal problem from different perspectives, namely the one proposed by Rivers [19, 20], based on a moving average filtering, and the one proposed by Münch, based on a combined wavelet-Fourier filtering [21].

4.3.4 Projection Despeckling

In each projection image few (about 0.5%) pixels with an abnormal number of counts, either lower or higher than the neighboring pixels, are present. Their appearance is not reproducible neither in space nor in time and their content cannot be correlated with the actual number of impinging photons. To remove these speckles, which cause streak artifacts in the reconstructed image, they first need to be recognized and then replaced. The procedure is based on a slightly different version of the alpha-trimmed filter described in the previous section, modified in order to filter only the bad pixels: for each pixel position i the average $\bar{f}(i)$ and standard deviation $\sigma(i)$ of the pixels comprised within a confidence window are computed, then the pixel of interest is replaced only if its value differs from the mean value more than $N\sigma(i)$, N being a parameter of the filter. In this way N acts as a filter sensitivity threshold, where if $N \to 0$ all the pixels are filtered, as in the implementation reported in Sect. 4.3.3, while if $N \to \infty$ no pixels are modified. Moreover, when calculating the average and standard deviation the $h - c$ smallest values and the $h - c$ largest values are excluded, meaning that pixels with either abnormally high or low counts can be easily discarded. For the projection despeckling, the filter window is a 5×5 pixels square and the confidence window is a 3×3 pixels square, while the optimization of the parameter N is reported in the results Sect. 4.4.

4.3.5 Around-Gap Equalization

The fifth and last step of the pre-processing is a dedicated procedure for equalizing the pixels around the gaps between modules. This further equalization is required since many adjacent columns of close-to-edge pixels show a sub-optimal gain behaviour. This effect involves a large number of pixel columns (30–40 columns across the gap),

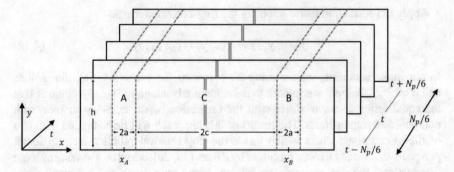

Fig. 4.4 Illustration of the equalization procedure: pixels of the projection t within the volume C are those to be equalized. See text for the complete description. The figure is not to scale

hence the action of the ring removal filter, which operates with a 10 pixels window, is not sufficient. This procedure is based on a moving average along the projection axis and it is described as follows:

- Given a projection t, a volume C of width $2c = 40$ pixels, height equal to the full height of the projection and depth $N_p/3$, where N_p is the number of projection, is selected across the gap between 2 modules. Other two volumes (A and B) with the same height, depth and a width of $2a = 10$ pixels are selected adjacent to C (see Fig. 4.4).
- The mean value along x and t axis is computed for the volumes A and B

$$\bar{f}_A(y;t) = \frac{1}{2aN_p/3} \sum_{x=x_A-a}^{x_A+a} \sum_{t'=t-N_p/6}^{t+N_p/6} f(x, y; t'),$$

$$\bar{f}_B(y;t) = \frac{1}{2bN_p/3} \sum_{x=x_B-b}^{x_B+b} \sum_{t'=t-N_p/6}^{t+N_p/6} f(x, y; t') \tag{4.13}$$

- The mean value along t is computed for the volume C

$$\bar{f}_C(x, y; t) = \frac{1}{N_p/3} \sum_{t'=t-N_p/6}^{t+N_p/6} f(x, y; t') \tag{4.14}$$

- The equalization factor is computed as

$$\text{eq}(x, y; t) = \frac{u(x)\bar{f}_A(y; t) + v(x)\bar{f}_B(y; t)}{\bar{f}_C(x, y; t)} \tag{4.15}$$

where the weights $u(x)$ and $v(x)$ are the normalized distances between the pixel within the gap C and the regions A and B, as defined in Sect. 4.3.2.

- The image is multiplied by the equalization factor

$$f_{\text{around}}(x, y; t) = f(x, y; t)\text{eq}(x, y; t) \tag{4.16}$$

In order for this procedure to be effectively used, the pixels within the regions A and B must not show a sub-optimal behavior. Moreover, as mentioned for the dynamic flat fielding and ring removal steps, the around-gap fixing equalization factor depends on the projection index, thus allowing to compensate for slow gain variations of close-to-gap pixels.

4.4   Pre-processing Procedure: Results

The effectiveness of the described procedure is tested on a breast surgical specimen with a diameter of 10 cm containing an infiltrating ductal carcinoma with a maximum dimension of 2.5 cm (sample B of Chap. 6). The sample is imaged at 32 keV and detector threshold set to 3 keV, delivering 20 mGy of mean glandular dose over 1200 equally spaced projections spanning an angle of 180 deg. The projections, either with or without the phase retrieval, are reconstructed via a standard FBP with Shepp-Logan filtering.

In order to compare the flat fielding procedures in the projection space, two sets of 1200 flat projections were acquired with different photon fluences: one is collected with a low photon fluence to simulate the sample attenuation, the other, acquired with a 4 times higher statistics, is used for the flat fielding. This choice is made to uncouple the effects of time and exposure on the detector's gain, thus having two datasets with the same acquisition time (i.e. acquired after the same time from the polarization of the CdTe sensor) but different exposures. In panels (a), (b) of Fig. 4.5, a detail of the first projection normalized with the static and the dynamic flat field approach is shown: at the center of both images a cluster of pixels with a gain lower than the neighboring ones is present. Observing the same region at a later time, it is evident that the cluster exhibits a gain variation which is more pronounced for the static flat fielding, in panel (c), with respect to the dynamic flat fielding, in panel (d). Focusing on the intensity plots as a function time, in panels (e)–(f), of a group of pixels within the cluster, it is clear that the gain variation of the statically flat fielded ($\sim$55%) dataset is significantly higher with respect to the one ($\sim$20%) of the dynamically flat fielded projections. Moreover, as it should be expected, the latter shows a smoother time-dependence which can be better compensated by the ring-removal procedure. The effects of each uncompensated crystal defect can be traced through the tomographic reconstruction process. In Fig. 4.6, panel (a), detail of the reconstructed image corresponding to a row through the defective pixel cluster obtained with the static flat fielding is shown: a bright streak-like artifact embedded within a partial ring artifact, due to the uncompensated gain variation, is observed. Panel (b) reports the same detail when the dynamic flat field approach is used: in this case the streak is barely visible while the ring has been removed. In both images

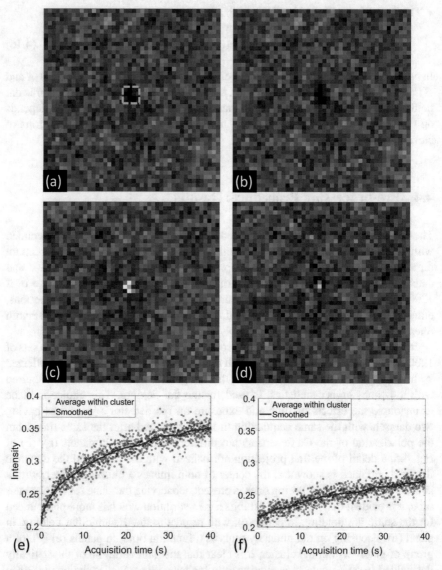

Fig. 4.5 Comparison between static and dynamic flat-fielding procedures in the projection space using two flat dataset with different statistics. In (**a**) and (**c**) the first and last projections when the static flat field is applied, in (**b**) and (**d**) the first and last projections when the dynamic flat field is applied. In (**e**) and (**f**) the average intensity of the bad pixel cluster (dashed line in (**a**)) as a function of time for the static and dynamic flat field respectively. Smoothed line is produced through moving average with a with a 100 points window

the whole pre-processing procedure has been applied in order to highlight only the effect of the flat fielding in the final reconstruction.

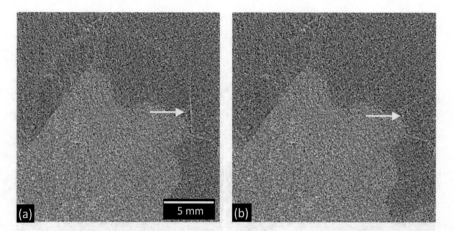

Fig. 4.6 Detail of a reconstruction obtained applying the static (**a**) and the dynamic (**b**) flat fielding. The arrow indicates a streak artifact clearly visible in (**a**) while it is barely visible in (**b**)

Panels (a), (c) in Fig. 4.7 show, respectively, the sinogram and the tomographic reconstruction of the sample where only the flat fielding has been applied. The sample was imaged using 4 modules of the detector, thus in the sinogram only 3 gaps are visible, producing marked ring artifacts in the reconstruction. The artifacts cover only half of circumference because the projections are acquired over 180 degrees. In panels (b), (d) both the sinogram and the reconstruction are reported after the gap seaming: given the small size of the gaps (3 pixels wide) the interpolation does not introduce significant artifacts, thus preserving the anatomical information. Nevertheless, the resulting image is still affected from the presence of several artifacts which need to be corrected.

Panels (a), (d) of Fig. 4.8 show the sinogram and the reconstruction where the Rivers ring-removal filter [19, 20] has been applied with a window width of 11 pixels, while in panels (b), (e) the Münch filter [21] has been applied with a decomposition level 5 and a width of the Gaussian bandpass function of 3. From the sinograms, it can be seen that neither the Rivers nor the Münch filter are optimal: in both cases most of the rings are only partially compensated resulting in arc (i.e. partial ring) artifacts. In particular, focusing on the Rivers approach where a constant equalization factor is used, the artifacts appear to be brighter at the top of the sinogram, well corrected in the central part and darker at the bottom. Again, this is due to the time gain variation which occurs to some pixels as previously described. The Münch filter yields slightly better results on the rings but it introduces a low spatial frequency modulation strongly affecting the image quality. Comparing these results with panels (c), (f), obtained with the procedure described in Sect. 4.3.3, it is clear that the latter yields the best results, substantially removing most of the ring artifacts. It is worth noticing that the main advantage of this approach is the presence of an equalization factor varying with the projection index.

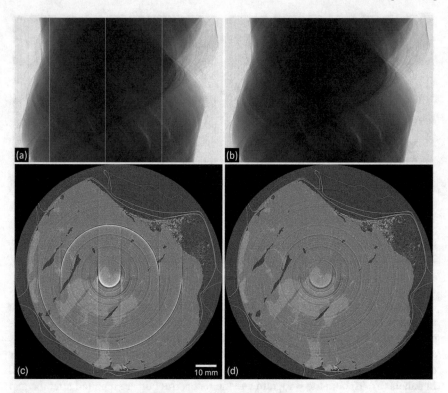

Fig. 4.7 Sinograms and reconstructions obtained before (**a**), (**c**) and after (**b**), (**d**) the gap seaming

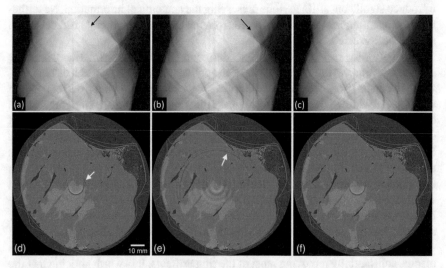

Fig. 4.8 Sinograms and reconstructions obtained applying Rivers (**a**), (**d**), Münch (**b**), (**e**) and dynamic (**c**), (**f**) ring removal filters. Sinograms are inverted and displayed on a logarithmic scale for better visualizing the action of the filters. The arrows in both the sinograms and the reconstructions indicate uncompensated ring artifacts

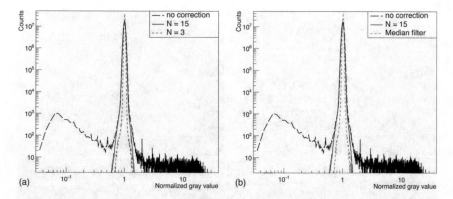

Fig. 4.9 Histograms for the despeckling filter optimization. In **a** the non filtered spectrum (long-dashed line) is compared with the filtered ones (solid line for N = 15, shirt-dashed line for N = 3), in (**b**) also the median filtered spectrum (shirt-dashed line) is reported

As reported in Sect. 4.3.4, the parameter N of the despeckling filter should be optimized in order to remove only the bad pixels. For this purpose a dataset of 1300 flat projections has been acquired and subdivided into two datasets consisting of the even and the odd projection, respectively. Then, the even projections were divided, pixel by pixel, by the odd projections. In this way the gain dependence from time and exposure is matched, and the distribution of the bad pixels alone can be studied. The gray level histogram of the resulting dataset is plotted in panel (a) of Fig. 4.9 (black dashed line): if no bad pixels are present, the distribution should be a Gaussian centered around one, whose width is only dependent on the photon statistics. On the contrary, the presence of bad pixels widens the distribution on both sides. The despeckling filter is expected to suppress the tails of the distribution without affecting the width of the Gaussian, i.e. the statistical noise. By varying continuously the filter parameter N it is found out that values around 15 satisfy this request (blue solid line) while, for lower N (e.g., N = 3, red dashed line), the statistical noise is reduced, meaning that a certain level of correlation among pixel is introduced and the image is unnecessarily smoothed. The same overcorrection effect is observed when applying common despeckling filters, such as the median filter, as reported in panel (b). Once the parameter N has been optimized, the despeckling filter can be applied to the projections. Panels (a), (b) in Fig. 4.10 show a detail of the sinogram before and after the application of the filter, respectively: the bad pixels have been removed without affecting the image noise and texture. The effect of the filter on the reconstruction is reported in panels (c), (d) of Fig. 4.10, where in the unfiltered image several striking artifacts due to bad pixels are visible. Here, it has to be remarked that the optimization of the parameter N is crucial since an excessive smoothing of the projections may disrupt the edge-enhancement effect, which is one of the key features of PB breast CT.

The last step of the pre-processing procedure is the around-gap equalization. In facts, referring to panel (a) in Fig 4.11, two wide rings corresponding to the regions

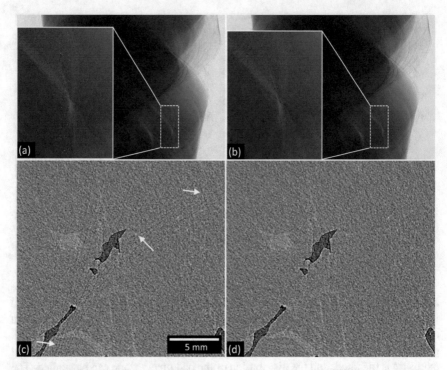

Fig. 4.10 Sinograms and reconstructions before (**a**), (**c**) and after (**b**), (**d**) and the application of the despeckling filter. Arrows indicate some of the streaks in the reconstruction

around the gaps between modules, can still be observed. Once the equalization procedure is applied the rings are removed and the final reconstructed image, shown in panel (b), is free from major artifacts.

After the projections have been pre-processed, the (two-materials, $(\delta_1 - \delta_2)/(\beta_1 - \beta_2) = 869$) phase-retrieval algorithm is applied. Noticeably, the phase-retrieval algorithms produces a remarkable increase in the contrast-to-noise-ratio, thus highlighting also uncompensated artifacts which may be barely visible in the phase-contrast images. In panels (a), (c), of Fig. 4.12 a detail of the reconstruction processed only with the first two steps of the pre-processing procedure (namely, flat fielding and gap seaming) is reported with and without phase retrieval: in both cases severe ring artifacts are observed but, when phase retrieval is applied, streak artifacts arising from uncompensated speckles become evident, definitely impairing the image quality. Conversely, when the whole pre-processing is applied, both images without and with phase-retrieval, in panels (b), (d), do not report significant artifacts. In this context, it should be stressed that the optimization of the pre-processing procedure must account also for the subsequent image processing (e.g., phase retrieval) in order to yield a high quality image. In Fig. 4.13 the final result of the data processing, comprising the pre-processing and the phase-retrieval procedure, is shown: the extension,

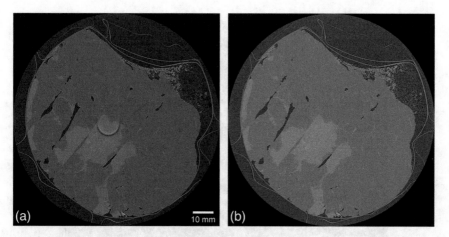

Fig. 4.11 Reconstructions before (**a**) and after (**b**) the around-gap equalization

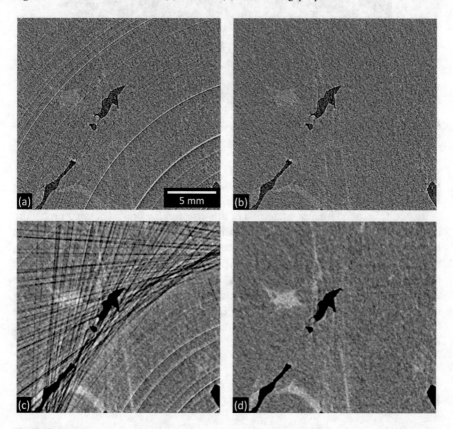

Fig. 4.12 Detail of a reconstruction without (**a**), (**b**) and with (**c**), (**d**) the phase retrieval. In (**a**), (**c**) only the flat fielding and gap seaming steps are applied, in (**b**), (**d**) the whole pre-processing procedure is used

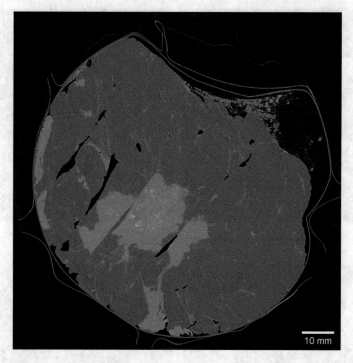

10 mm

Fig. 4.13 Final reconstruction obtained subsequently applying the pre-processing procedure and the phase-retrieval

shape and boundaries of both the tumoral and glandular tissue (light gray) embedded in the adipose background (dark gray) are clearly visible without artifacts.

In addition to the images presented in this chapter, the pre-processing procedure has been successfully applied to a great variety of breast-like samples, spanning from test objects to a number of surgical specimens, within a wide range of beam energies, fluences and detector thresholds [22–26]. As a general remark, it is the author's belief that high-Z single-photon-counting detectors will be widely used in future CT applications, especially in medical imaging, due to their high-efficiency, low noise and spectral performances: in this framework, the pre-processing procedure presented in this chapter may represent a useful scheme to be extended to other imaging contexts.

References

1. Brombal L, Donato S, Brun F, Delogu P, Fanti V, Oliva P, Rigon L, Di Trapani V, Longo R, Golosio B (2018) Large-area single-photon-counting cdte detector for synchrotron radiation computed tomography: a dedicated pre-processing procedure. J Synchrotron Radiat 25(4). https://doi.org/10.1107/S1600577518006197

2. Vedantham S, Shrestha S, Karellas A, Shi L, Gounis MJ, Bellazzini R, Spandre G, Brez A, Minuti M (2016) Photon-counting hexagonal pixel array CdTe detector: Spatial resolution characteristics for image-guided interventional applications. Med Phys 43(5):2118–2130. https://doi.org/10.1118/1.4944868

3. Ballabriga R, Alozy J, Campbell M, Frojdh E, Heijne E, Koenig T, Llopart X, Marchal J, Pennicard D, Poikela T et al (2016) Review of hybrid pixel detector readout ASICs for spectroscopic X-ray imaging. J Instrum 11(01):P01007

4. Takahashi T, Watanabe S (2001) Recent progress in CdTe and CdZnTe detectors. IEEE Trans Nucl Sci 48(4):950–959. https://doi.org/10.1109/23.958705

5. Taguchi K, Iwanczyk JS (2013) Vision 20/20: Single photon counting X-ray detectors in medical imaging. Medical Phys 40(10). https://doi.org/10.1118/1.4820371

6. Delogu P, Golosio B, Fedon C, Arfelli F, Bellazzini R, Brez A, Brun F, Di Lillo F, Dreossi D, Mettivier G et al (2017b) Imaging study of a phase-sensitive breast-CT system in continuous acquisition mode. J Instrum 12(01):C01016. https://doi.org/10.1088/1748-0221/12/01/C01016

7. Mozzanica A, Bergamaschi A, Brueckner M, Cartier S, Dinapoli R, Greiffenberg D, Jungmann-Smith J, Maliakal D, Mezza D, Ramilli M et al (2016) Characterization results of the JUNGFRAU full scale readout ASIC. J Instrum 11(02):C02047. https://doi.org/10.1088/1748-0221/11/01/P01007

8. Astromskas V, Gimenez EN, Lohstroh A, Tartoni N (2016) Evaluation of polarization effects of e- collection Schottky CdTe Medipix3RX hybrid pixel detector. IEEE Trans Nucl Sci 63(1):252–258. https://doi.org/10.1109/tns.2016.2516827

9. Park SE, Kim, JG Hegazy M, Cho MH, Lee SY (2014) A flat-field correction method for photon-counting-detector-based micro-CT. Proc SPIE 9033:90335. https://doi.org/10.1117/12.2043317

10. Pennicard D, Graafsma H (2011) Simulated performance of high-Z detectors with medipix3 readout. J Instrum 6(06):P06007. https://doi.org/10.1088/1748-0221/6/06/P06007

11. Knoll GF (2010) Radiation detection and measurement. Wiley. https://doi.org/10.1016/S0969-806X(00)00323-6

12. Delogu P, Brombal L, Di Trapani V, Donato S, Bottigli U, Dreossi D, Golosio B, Oliva P, Rigon L, Longo R (2017a) Optimization of the equalization procedure for a single-photon counting cdte detector used for ct. J Instrum 12(11):C11014. https://doi.org/10.1088/1748-0221/12/11/C11014

13. Bellazzini R, Spandre G, Brez A, Minuti M, Pinchera M, Mozzo P (2013) Chromatic X-ray imaging with a fine pitch cdte sensor coupled to a large area photon counting pixel asic. J Instrum 8(02):C02028. https://doi.org/10.1088/1748-0221/8/02/C02028

14. Delogu P, Oliva P, Bellazzini R, Brez A, De Ruvo P, Minuti M, Pinchera M, Spandre G, Vincenzi A (2016) Characterization of pixirad-1 photon counting detector for X-ray imaging. J Instrum 11(01):P01015. https://doi.org/10.1088/1748-0221/11/01/P01015

15. Brun F, Di Trapani V, Dreossi D, Rigon L, Longo R, Delogu P (2019) Towards in vivo k-edge X-ray micro-CT with the pixirad-i/pixie-iii detector. In: World congress on medical physics and biomedical engineering, pp 123–126. Springer. https://doi.org/10.1007/978-981-10-9035-6_22

16. Van Nieuwenhove V, De Beenhouwer J, De Carlo F, Mancini L, Marone F, Sijbers J (2015) Dynamic intensity normalization using eigen flat fields in X-ray imaging. Opt Express 23(21):27975–27989. https://doi.org/10.1364/OE.23.027975

17. Brun F, Delogu P, Longo R, Dreossi D, Rigon L (2017) Inpainting approaches to fill in detector gaps in phase contrast computed tomography. Measur Sci Technol 29(1):014001. https://doi.org/10.1088/1361-6501/aa91ad

18. Bednar J, Watt T (1984) Alpha-trimmed means and their relationship to median filters. IEEE Trans Acoust Speech Signal Process 32(1):145–153

19. Rivers M (1998) Tutorial introduction to X-ray computed microtomography data processing. http://www.mcs.anl.gov/research/projects/X-raycmt/rivers/tutorial.html

20. Boin M, Haibel A (2006) Compensation of ring artefacts in synchrotron tomographic images. Opt Express 14(25):12071–12075
21. Münch B, Trtik P, Marone F, Stampanoni M (2009) Stripe and ring artifact removal with combined wavelet—Fourier filtering. Opt Express 17(10):8567–8591. https://doi.org/10.1364/OE.17.008567
22. Contillo A, Veronese A, Brombal L, Donato S, Rigon L, Taibi A, Tromba G, Longo R, Arfelli F (2018) A proposal for a quality control protocol in breast CT with synchrotron radiation. Radiol Oncol 52(3):1–8. https://doi.org/10.2478/raon-2018-0015
23. Brombal L, Donato S, Dreossi D, Arfelli F, Bonazza D, Contillo A, Delogu P, Di Trapani V, Golosio B, Mettivier G et al (2018) Phase-contrast breast CT: the effect of propagation distance. Phys Med Biol 63(24):24NT03. https://doi.org/10.1088/1361-6560/aaf2e1
24. Brombal L, Golosio B, Arfelli F, Bonazza D, Contillo A, Delogu P, Donato S, Mettivier G, Oliva P, Rigon L et al (2018c) Monochromatic breast computed tomography with synchrotron radiation: phase-contrast and phase-retrieved image comparison and full-volume reconstruction. J Med Imaging 6(3):031402. https://doi.org/10.1117/1.JMI.6.3.031402
25. Brombal L, Golosio B, Arfelli F, Bonazza D, Contillo A, Delogu P, Donato S, Mettivier G, Oliva P, Rigon L et al (2018) Monochromatic breast ct: absorption and phase-retrieved images. In: Medical imaging 2018: physics of medical imaging, vol 10573, page 1057320. International Society for Optics and Photonics. https://doi.org/10.1117/12.2293088
26. Donato S, Brombal, L Tromba G, Longo R et al (2019) Phase-contrast breast-CT: optimization of experimental parameters and reconstruction algorithms. In: World congress on medical physics and biomedical engineering 2018, pp 109–115. Springer. https://doi.org/10.1007/978-981-10-9035-6_20

Chapter 5
Experimental Optimization of Propagation-Based BCT

Effective design and implementation of a propagation-based CT setup require careful optimization both in terms of physical parameters (hardware) and data processing (software). The goal of the present chapter is to describe and provide a scientific justification for several of these aspects, combining a theoretical/mathematical background with experimental results in the context of the SYRMA-3D project. Specifically, in the first two sections a model describing the propagation of signal and noise through the imaging chain will be introduced; by comparing experimental data with theoretical predictions, the effects of propagation distance and detector pixel size on image noise and signal-to-noise-ratio will be discussed and the consequences of these findings on the SYRMEP beamline upgrade will be presented. In the third section the development of a beam filtration system to produce a vertically wider and more uniform X-ray intensity distribution at the sample position will be described. In the fourth and last section a post-reconstruction phase-retrieval pipeline, aiming at compensating for periodic artifacts arising in multi-stage CT acquisitions, will be introduced, also providing a mathematical proof of the equivalence to its pre-reconstruction counterpart.

5.1 The Effect of Propagation Distance

As PB imaging relies on the free-space propagation of the perturbed X-ray wavefront between the object and the detector, it is not surprising that the object-to-detector (or propagation) distance plays a crucial role in determining the final image appearance, as already mentioned in Sect. 2.2. For this reason, a formal model describing the effect of propagation distance on image quality metrics as noise, signal-to-noise ratio (SNR) and spatial resolution is introduced in this section and applied to the specific case of

L. Brombal, *X-Ray Phase-Contrast Tomography*, Springer Theses, https://doi.org/10.1007/978-3-030-60433-2_5

PBBCT. The model is mainly derived from theoretical works by Iakov I. Nesterets and Timur Gureyev [1, 2]. Comparisons between theory and experimental results will be shown in the next sections, and the impact of the findings on the upgrade of the experimental setup will be discussed.

5.1.1 Theoretical Model

In Sect. 2.4 the propagation process has been described as an operator acting on the X-ray intensity distribution emerging from the object at the object plane. To take into account the realistic case of a divergent beam, where the geometrical magnification M is not negligible, the (forward) propagation operator previously introduced has to be slightly modified to

$$H' = \left(1 - \frac{z_1}{M}\frac{\delta}{2k\beta}\nabla_{xy}^2\right) = \left(1 - \frac{z'\lambda\delta}{4\pi\beta}\nabla_{xy}^2\right) \tag{5.1}$$

where $z' = z_1/M$ is referred to as effective propagation distance and the definition $k = 2\pi/\lambda$ has been inserted. This equation implies that phase-contrast signal appears at the interfaces of/within the imaged object, where the intensity Laplacian is expected to be significantly different from zero, and it is proportional to the propagation distance. On the contrary, within uniform regions of the collected image (i.e. far from sharp details) the Laplacian term can be neglected and the detected signal only depends on the attenuation properties of the object. For this reason, if measured far from sharp interfaces where phase-contrast is present, neither the (large-area) signal nor the statistical noise (that, in case of Poissonian statistics, is proportional to the square root of the signal) are affected by the propagation process. The same consideration holds also for the image (large-area) contrast, that is defined as the difference between a detail and background signals, measured far from interfaces, normalized to the background.

5.1.1.1 Effects on Spatial Resolution

As stated previously, while not affecting large-area signal and noise, propagation produces the edge-enhancement effect which boosts the high spatial frequency component of the image, hence improving the spatial resolution. To understand this effect quantitatively, let us define the image blur in a planar image, which is inversely proportional to the spatial resolution, as the standard deviation of the detector PSF (here the source is assumed to be point-like) considering, for the sake of simplicity, the mono-dimensional case

$$(\Delta x)\,[\mathrm{PSF_{det}}] = \left(\int x^2 \mathrm{PSF_{det}}(x)\,\mathrm{d}x \right)^{1/2} \tag{5.2}$$

Starting from this definition it can be demonstrated that, by applying the propagation operator introduced in Eq. (5.1), the effective detector PSF width is decreased in the propagation process, i.e. the spatial resolution is improved [2]:

$$(\Delta x)^2_{z'}\,[\mathrm{PSF_{det}}] = (\Delta x)^2\,[\mathrm{PSF_{det}}] - \frac{z'\lambda\delta}{2\pi\beta} \tag{5.3}$$

Of note, the last term of the equation, determining the narrowing of the effective PSF, depends linearly on the effective propagation distance.

The propagation process is followed by the application of the phase-retrieval algorithm. As described in Sect. 2.4, the PhR is a low-pass filter, thus it affects the image by reducing noise and degrading the spatial resolution. Following the same line of reasoning used to describe the change in resolution due to the propagation, and recalling that the PhR operator is the inverse of the propagation operator, the application of PhR leads to an increase of the image blur that reads:

$$(\Delta x)^2_{\mathrm{PhR}}\,[\mathrm{PSF_{det}}] = (\Delta x)^2\,[\mathrm{PSF_{det}}] + \frac{z'\lambda\delta}{2\pi\beta} \tag{5.4}$$

where the term responsible for the PSF widening is the same as in Eq. (5.3) but with opposite sign. At this point it is clear that combining propagation and phase retrieval means to add and subtract the term $z'\lambda\delta/(2\pi\beta)$, thus leaving the spatial resolution unaltered. Since the latter statement is valid for each planar projection image, it is trivial to conclude that it applies also to the reconstructed tomographic volume.

5.1.1.2   Effects on Image Noise

Having shown that the effects of phase retrieval and forward propagation on spatial resolution exactly compensate each other, let us steer the attention on the effect of phase retrieval on CT image noise. A rather general model of image noise in reconstructed CT slices has been recently introduced by [3] and its formulation is well suited to include analytically the phase-retrieval filter.

According to the model, by assuming a Poisson dominated detector noise, flat-fielded bi-dimensional projection images, stable source intensity and imaging setup, and parallel beam tomographic reconstruction performed through the Filtered-Back-Projection (FBP) algorithm, the variance (var) in homogeneous region of a CT image is given by

$$\mathrm{var} = \frac{f(A; d/h)\,F_{\mathrm{obj}}}{N_p h'^4 \Phi \mathrm{DQE_0} T_{\mathrm{obj\text{-}det}}} \tag{5.5}$$

where F_{obj} accounts for X-rays attenuation in the object, N_p is the number of projections in the tomographic scan, Φ is the X-ray fluence at the object (in number of photons per square millimeter), DQE_0 is the detector quantum efficiency at zero spatial frequency, $T_{obj\text{-}det}$ is the transmittance through the object-to-detector distance (usually transmission in air) and $h' = h/M$ is the effective pixel size accounting for the geometrical magnification M and it is assumed to be bi-dimensional with equal width and height. The dimensionless function $f(A; d/h)$ accounts for the tomographic process, the detector response and phase retrieval, and it is written as:

$$f(A; d/h) = 2\pi^2 \int_0^{\frac{1}{2}} dU\, G^2(U) F_{interp}(U) \int_{-\frac{1}{2}}^{\frac{1}{2}} dV \frac{MTF^2(U, V; d/h)}{\left[1 + 4A\left(U^2 + V^2\right)\right]^2} \quad (5.6)$$

Here $G(U)$ is the the CT filter, $F_{interp}(U)$ describes the effect on noise of the interpolation from polar to Cartesian coordinates in the backprojection process, $MTF(U, V; d/h)$ is the detector planar modulation transfer function parametrized through the dimensionless quantity d/h, where d is the full width at half maximum (FWHM) of the detector's point spread function. Of note, the integration variables U and V are dimensionless normalized frequencies expressing the fraction of twice the maximum detected frequency (Nyquist frequency), hence fractions of $(h/M)^{-1}$. Finally, the dimensionless parameter A depends on the refractive properties of the sample, on the setup geometry and on the detector pixel size as

$$A = (\pi/4) \frac{z'\lambda\delta}{h'^2\beta} \quad (5.7)$$

where δ/β can be referred to both single- and two-materials phase retrieval (see Sect. 2.5).

Despite its rather complex formulation, the function f is key in understanding the effect of phase retrieval on image noise which is summarized in the denominator of Eq. (5.6). In facts, when no PhR is applied $A = 0$ and, as a consequence, the function f does not explicitly depend neither on the effective propagation distance nor on the effective pixel size: in this case, following Eq. (5.5), the image noise ($\sigma = \sqrt{var}$) at fixed sample fluence is found to be proportional to $1/h'^2 = M^2/h^2$, which is a known result in the context of conventional CT [4]. On the contrary, if PhR is applied $A > 0$ and the denominator in Eq. (5.6) is larger than 1, hence the function f gets smaller if compared with the case $A = 0$, bringing to a reduction in image noise. More in detail, an increase of propagation distance and/or a decrease on the effective pixel size, bring to an increase of the parameter A which, in turn, determines a decrease in image noise.

If we define σ_{PhR} and σ_{noPhR} to be the noise in a flat region of a tomographic image obtained with and without the application of PhR, respectively, the noise reduction factor associated to PhR can be written as

$$\frac{\sigma_{\text{PhR}}}{\sigma_{\text{noPhR}}} = \left[\frac{f(A; d/h)}{f(0; d/h)} \right]^{1/2} \tag{5.8}$$

In general, this factor cannot be calculated analytically as to compute the function f considering a realistic MTF, reconstruction filters and interpolations numerical integration is required: these realistic parameters will be introduced in the next section. Anyway, following the work by Nesterets and colleagues [1], an explicit analytical formula can be found by assuming a flat detector MTF up to the Nyquist frequency, a ramp tomographic filter, the use of nearest neighbour interpolation and large values of A ($A \gg 1$):

$$\frac{\sigma_{\text{PhR}}}{\sigma_{\text{noPhR}}} = \left[\frac{3\pi}{8} \frac{\ln A - 1}{A^2} \right]^{1/2} \tag{5.9}$$

Of course, considering the simplifications introduced in describing both the detector and the tomographic reconstruction process, this equation has to be regarded as a first approximation providing a rough estimate of the noise-reduction factor. On the other hand, in the specific case of the SYRMA-3D BCT project, the assumption $A \gg 1$ is rather reasonable since $z' > 1$ m, $h' \sim 50$ μm, $\delta/\beta \sim 10^3$ and $\lambda \sim 3 \cdot 10^{-11}$ m, yielding $A \gtrsim 10$.

At this point, recalling that large-area signal is not altered by the application of the PhR, as shown in Fig. 5.1, and by defining the signal-to-noise-ratio in a tomographic image as $SNR = \langle I \rangle / \sigma$, where $\langle I \rangle$ denotes the image mean value in a region far from sharp interfaces, the SNR gain factor due to the phase retrieval will be:

$$SNR_{\text{gain}} = \frac{SNR_{\text{PhR}}}{SNR_{\text{noPhR}}} = \frac{\langle I \rangle / \sigma_{\text{PhR}}}{\langle I \rangle / \sigma_{\text{noPhR}}} = \left[\frac{8}{3\pi} \frac{A^2}{\ln A - 1} \right]^{1/2} \tag{5.10}$$

This equation represents a crucial result since it allows to determine the effect of all the experimental parameters, summarized by A, on the image SNR and, ultimately, on the visibility of details. Assuming that the logarithmic term varies slowly, the SNR gain increases almost linearly with the parameter A. By recalling its definition in Eq. (5.7), this means that the gain factor scales approximately linearly with the propagation distance. Considering the realistic parameters described above, the expected SNR gain is between 1 and 2 orders of magnitude, which means that phase retrieval has a dramatic impact on the image quality. A convincing experimental demonstration of this effect, based on images of rabbit kitten lungs, can be found in [5].

The effects of propagation, phase retrieval and their combination on the tomographic image signal, noise, SNR and blur are schematically summarized in Table 5.1.

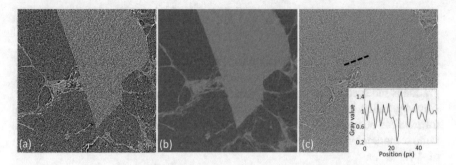

Fig. 5.1 Detail of a breast specimen PB tomographic image reconstructed without (**a**) and with (**b**) PhR. In (**c**) the ratio of (**b**) and (**a**) is reported: the application of PhR does not alter the image signal except for sharp interfaces where edge-enhancement effect is present as shown by the inset in (**c**) reporting the intensity profile along the black dashed line

Table 5.1 Schematic summary of the effects of propagation and phase retrieval on common image quality metrics. Arrows identify whether the image quality increases (green), decreases (red) or remains constant (black)

	Contrast	Noise	SNR	Blur (or resolution^{-1})
Propagation	↔	↔	↔	↓
PhR	↔	↓	↑	↑
Propagation + PhR	↔	↓	↑	↔

5.1.2 Acquisition Parameters and Image Analysis

All the acquisitions performed to test the aforementioned model have been carried out at a fixed source-to-detector distance of 31.6 m, at a beam energy of 30 keV, and by positioning the sample at 3 different object-to-detector distances, 1.6, 3 and 9 m, respectively. These sample positions correspond to geometrical magnifications $M = [1.05, 1.10, 1.40]$, and to effective propagation distances $z' = z_1/M = [1.52\,\text{m}, 2.73\,\text{m}, 6.43\,\text{m}]$. As a general remark it is worth noting that, especially at high magnifications, the actual finite dimension of the source should be taken into account since it contributes to the overall image blurring, thus reducing the spatial resolution [6], as discussed in Sect. 2.3. Anyway, considering the actual source size, which is in the order of 100 μm [7], and the small magnification factors used (1.4 or lower), the finite source size effect can be (as a first approximation) neglected since, following Eq. (2.15), its contribution is smaller than the pixel size (60 μm).

Each scan is performed in 40 s, collecting 1200 projections over 180 deg with a rotation speed of 4.5 deg s^{-1}. The fluence on the detector plane was fixed in order to deliver a total mean glandular dose of 25 mGy at the patient position, i.e. 1.6 m object-to-detector distance. It should be noted that at larger distances both the effects of magnification and X-ray attenuation in air are not negligible and they determine a higher delivered dose. In particular, air attenuation produces a dose increase of ~10% at 3 m and ~45% higher at 9 m of propagation at 30 keV: considering in-vivo applications, this issue can be overcome by positioning a vacuum pipe between the object and the detector, thus avoiding air attenuation. Anyway, as it will be clear in the next section, it can be argued that both magnification and air attenuation effects are largely compensated by the SNR increase at larger distances, leaving room for the possibility of a major dose reduction.

The scanned sample is a portion of a total breast mastectomy containing an epithelial and stromal sarcomatoid carcinoma. After the formalin fixation and sealing in a vacuum bag, the sample diameter is of about 12 cm. The projection images are pre-processed as described in Chap. 4 and phase retrieved with $(\delta_1 - \delta_2)/(\beta_1 - \beta_2) = 795$ (two-materials PhR), corresponding to a glandular/adipose interface, according to the values extracted from a publicly available database [8]. CT images are reconstructed via a parallel-beam FBP with a Shepp-Logan filter, meaning that, in the model introduced in the previous section (see Eq. (5.6)), $G(U) = U \operatorname{sinc}(U)$ where $\operatorname{sinc}(U) = \sin(\pi x)/(\pi x)$ is the normalized sinc function. The backprojection algorithm makes use of linear interpolation, therefore $F_{\text{interp}}(U) = [2 + \cos(2\pi U)]/3$. The detector MTF is modelled as a bi-dimensional sinc function, $MTF(U, V; 1) = \operatorname{sinc}(U)\operatorname{sinc}(V)$, which implies a bi-dimensional box-shaped point-spread-function (PSF) in real space having a width corresponding to the pixel size. The latter assumption, despite being an approximation, is rather reasonable for photon-counting detectors as Pixirad-8, where the PSF width is dominated by the physical pixel dimension, hence $d/h \simeq 1$.

As a first step of image analysis, the SNR of the images prior to the phase retrieval is measured within circular ROIs (4000 pixels each) embedded in the tumoral tissue, avoiding sharp edges. Following the model introduced in the previous section, if no phase-retrieval is applied SNR should not change significantly when the propagation distance is varied, being equal to the SNR that would be observed in the contact (i.e. object) plane, except for magnification effects.

Specifically, the SNR measured from experimental images is defined as:

$$\text{SNR} = \frac{\langle I \rangle}{\sigma} \frac{M}{M_0} \sqrt{\frac{N_0}{N}} \tag{5.11}$$

where $\langle I \rangle$ is the mean pixel value, σ the standard deviation in the ROI. To compensate for geometrical magnification, SNR is normalized to the magnification M over a reference value $M_0 = 1.05$, corresponding to the patient support position (effective propagation distance of 1.52 m). A detailed justification for this normalization factor is provided in Appendix B. Moreover, to make up for small fluence variations in different acquisitions, SNR is also normalized to the square root of the average number of counts in the detector N over the reference number N_0 corresponding to the recorded counts at 1.52 m of propagation. Of note, both normalization factors are rather small numbers (their product ranges from 1 to 1.4) compared to the SNR gain due to phase retrieval. The error associated to the SNR is given by the standard deviation of five SNR measurements performed in non-overlapping ROIs. SNR measurement is repeated on phase-retrieved images and SNR gain factor is calculated: it should be stressed that, while the introduced normalization factors play a role in calculating SNR, they are completely irrelevant for the calculation of the SNR gain factor since they cancel out as it is clear from Eq. (5.10).

Subsequently, the image contrast is measured from ROI pairs positioned both within tumor (subscript 1) and adipose (subscript 2) regions:

$$C = \frac{\langle I_1 \rangle - \langle I_2 \rangle}{\langle I_2 \rangle} \times 100 \qquad (5.12)$$

Since phase retrieval affects image noise while propagation affects spatial resolution, the contrast should not change neither with the application of the phase retrieval, nor varying the propagation distance. As for the SNR, the error associated to the contrast is given by the standard deviation of five contrast values measured in non-overlapping ROI pairs.

The spatial resolution is measured for the phase-retrieved images by selecting, for each distance, three line profiles across a sharp fat/tumor interface produced by a surgical cut. The line profiles are fitted with an error function (erf) and the FWHM of its derivative is measured. The spatial resolution is evaluated as the mean value of the three FWHMs and the error is estimated to be the maximum fluctuation around the mean value. According to the theory, excluding the effect of the magnification, the spatial resolution after the PhR should not vary at different propagation distances since, for each distance, the PhR is expected to produce the same resolution that would have been measured in the contact plane image. In order to consider only the intrinsic system's spatial resolution, the FWHM is measured in number of pixels instead of an absolute length.

5.1.3 Experimental Results

Many experimental results reported in this section are reproduced from [9] by permission of IOP Publishing.

In Fig. 5.2 the reconstructed slices at all effective propagation distances (1.52 m, 2.72 m, 6.44 m) without (a)–(c) and with (d)–(f) PhR are shown. With the aim of a visualization allowing a straightforward comparison between images with and without phase retrieval, the gray levels of all the images have been scaled by a normalization factor such that the average value of fibroglandular tissue far from interfaces is 1 while air is 0. Since tissue relaxation occurred and sample repositioning was needed, some morphological changes (e.g., different position of air gaps within the tissue) are observed at different propagation distances. Care was taken to ensure the best match at all distances in the region enclosed by the dashed line of panel (a), where all the measurements are performed. From the images it can be qualitatively

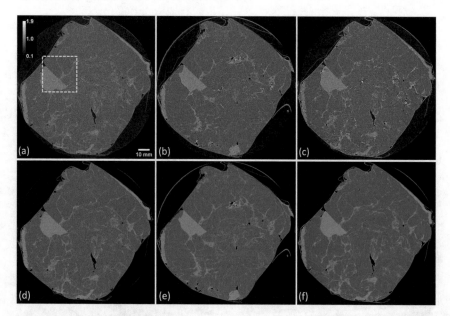

Fig. 5.2 Reconstructed slice acquired at effective propagation distances of 1.52 m (**a**), (**d**), 2.72 m (**b**), (**e**) and 6.44 m (**c**), (**f**). Images in the first row (**a**)–(**c**) are reconstructed without PhR, images in the second row (**d**)–(**f**) with PhR. The dashed square in (**a**) is the zoom region reported in Fig. 5.3. After the normalization described in text, images are displayed in a gray scale window ranging from 0 to 2, where 0 is a typical value of air and 1 a typical value of fibroglandular tissue. Morphological variations at different distances are due to sample repositioning and tissue relaxation within the sample holder

noted that, if no PhR is applied, no major variation in signal and noise is observed by varying the propagation distance, except for the sharp interfaces between adipose (dark gray) and tumor or fibroglandular (bright gray) tissue. On the contrary, when increasing propagation distances, the phase-retrieved reconstructions are smoother while no differences at tissues interfaces are observed.

The same effect is reported in a finer detail in Fig. 5.3, where a zoom on a sharp adipose/tumor interface produced by a surgical cut is displayed. Considering the non-phase-retrieved images (a)–(c) it is clear that the edge-enhancement effect at the interfaces between the two different tissues is amplified at larger propagation distances, i.e. the high-spatial frequencies are boosted. This can be better visualized in panels (g)–(i) reporting the line intensity profiles of the non-phase-retrieved images. Besides the edge-enhancement effect, clearly visible in panel (i), the profiles show a high level of noise, possibly hampering tissue differentiation. On the other hand, when the PhR is applied (d)–(f), the edge appearance does not change by varying the propagation distance and the edge-enhancement is no longer present. Considering the respective line profiles reported in panels (j)–(l), a similar edge sharpness is observed at all distances and, when compared with the non-phase-retrieved images profiles, the noise level is significantly lower.

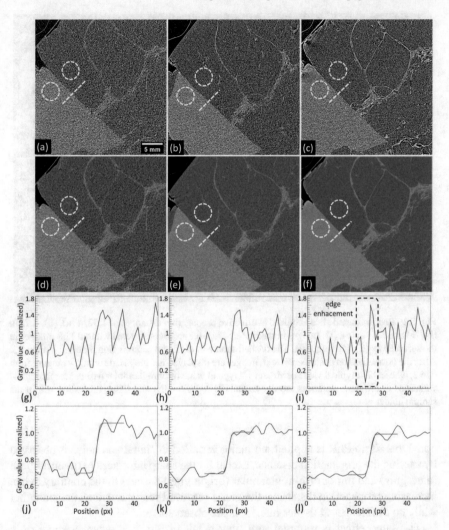

Fig. 5.3 Zoomed detail of Fig. 5.2 without (**a**)–(**c**) and with (**d**)–(**f**) phase retrieval at increasing propagation distances (from left to right). In panels (**g**)–(**i**) profiles obtained from the dashed lines in (**a**)–(**c**) are reported. In panels (**j**)–(**l**) profiles obtained from the dashed lines in (**d**)–(**f**) are reported along with the *erf* fit (red curve). In (**a**)–(**f**) one of the five pairs of circular ROIs used to determine contrast and SNR is displayed as an example

The quantitative results of the image analysis are reported in Table 5.2. As predicted by the theory (see Table 5.1) the SNR, calculated according to Eq. 5.11, does not vary significantly with the propagation distance if no PhR is applied, while its increase due to PhR is greater than a factor of 20 when considering 6.44 m of propagation distance. In addition, it must be noted that only little contrast variations (below 6%) are observed when changing the distance while, at a given position,

Table 5.2 Quantitative results. The uncertainty associated to each measure is enclosed between round brackets

	z'			
	PhR	1.52 m	2.72 m	6.44 m
SNR	No	1.63 (0.02)	1.63 (0.03)	1.62 (0.01)
	Yes	8.45 (0.13)	13.3 (0.3)	33.8 (0.7)
Contrast (%)	No	48.9 (0.5)	44.1 (0.5)	50.0 (0.4)
	Yes	48.6 (0.3)	44.2 (0.1)	49.1 ($<$ 0.1)
FWHM (px)	Yes	2.1 (0.5)	2.3 (0.3)	2.4 (0.2)

no relevant contrast alterations are associated to the PhR algorithm whose action is limited to image noise. The latter observation is of great importance in sight of the clinical application of this technique, since the image appearance will look 'familiar' to the clinician's eye, who will not require a specific training to read the images, as it may occur for other phase-contrast techniques. Furthermore, considering phase-retrieved images, the FWHM measured in pixel units does not vary significantly with the propagation distances and, in all cases, it was found to be slightly higher than 2 pixels (120 μm on the detector plane). This implies that, taking into account the magnification, the actual spatial resolution slightly improves at longer distances (FWHM 100 μm) at the expense of a smaller field of view.

With the aim of a better data visualization, the measured SNR gain, contrast and spatial resolution concerning the phase-retrieved images (points) and the theoretical predictions (lines) are plotted as a function of the propagation distance in Fig. 5.4. From the top panel it can be seen that the measured SNR gain is in remarkable agreement with the model results obtained via numerical integration considering realistic detector and reconstruction parameters (solid line). Interestingly, if the analytical formula given in Eq. (5.10) is followed instead of numerical integration, the predicted SNR gain factor (dashed line) is about 2-fold higher than the measured one. This can be easily explained taking into account the number of simplifications made in deriving that expression, the fundamental one being the rather unrealistic assumption of a detector featuring a constant MTF up to the Nyquist frequency: for this reason the values predicted according to the analytical formula constitute, in practice, an upper limit in terms of SNR gain when compared with experimental data. At the same time, it is worth mentioning that the factor of 2 difference between the two different approaches is almost constant at all the propagation distances, hence, even if Eq. (5.10) does not provide an accurate estimate of SNR gain factors in absolute terms, it still provides the correct trend with respect to the propagation distance. In addition, when comparing phase-retrieved images, a 4-fold increase in SNR is observed at 6.44 m with respect to the shortest propagation distance (1.52 m): remarkably, at a fixed propagation distance, such SNR increase would correspond to a 16-fold higher radiation dose.

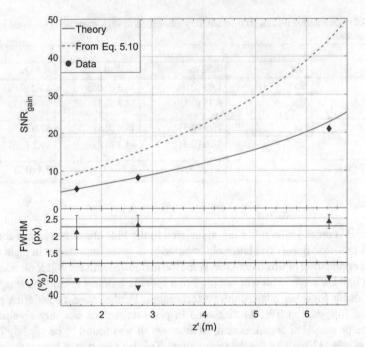

Fig. 5.4 Comparison between experimental results (points) and theoretical predictions (lines) as a function of the propagation distance. In the top panel the SNR gain factors calculated by using the analytical expression in Eq. (5.10) (dashed line) and by numerical integration from Eq. (5.8) with realistic parameters (solid line) are reported. Some error bars are smaller than points

As a final remark it is worth to mention that the model introduced and tested throughout this section is valid within the near-field propagation description or, equivalently, for large Fresnel numbers $N_F \gg 1$. Since the Fresnel number is inversely proportional to the propagation distance, this condition practically limits the maximum achievable SNR gain, which cannot arbitrarily increase. Of note, the requirement of a large Fresnel number is often relaxed in experimental practice and the near-field description is adopted even when $N_F \gtrsim 1$.

5.1.4   Consequences on the SYRMEP Upgrade

Improving any radiographic technique means either to provide a higher image quality at a constant dose or, equivalently, to provide the same image quality at a lower dose. In light of the results of the previous section, a longer propagation distance has the potential to dramatically improve PBBCT. Unfortunately, at the SYRMEP beamline, the patient support is at a fixed distance (30 m) from the source. This means that larger propagation distances can be only reached by further distancing

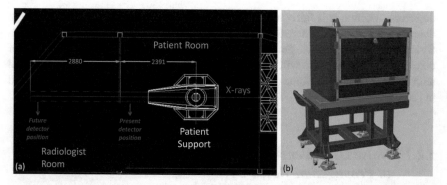

Fig. 5.5 Drawing of the patient room and the adjacent room, referred to as 'radiologist room' (**a**): dashed blue lines represent the propagation distance extension across the two rooms, allowing a gain of about 3 m in propagation distance (distances in figure are reported in mm). Drawing of the cabinet required to install the detector in the new position (**b**). The cabinet will be interlocked with the safety system constituting, in practice, an extension of the patient room

the detector from the support along the beam direction. This requires some major modifications to the present configuration of the beamline, where the maximum available sample-to-detector distance in the patient room is 1.6 m. As a consequence of the presented results, the realization of an *ad-hoc* designed extension beyond the patient room, depicted in Fig. 5.5, panel (a), has been funded. The extension will bring to a gain of about 3 m of propagation, corresponding to an object-to-detector distance of about 4.5 m. Due to radiation protection requirements, the detector will be enclosed within a dedicated cabinet, shown in panel (b), which will be interlocked with the safety system. According to the presented noise model, validated through experimental results hereby reported, this new configuration is expected to produce a SNR improvement of a factor of 2 or more with respect to the present setup at a constant fluence on the sample plane, i.e. at a constant delivered dose, thus constituting an major improvement in the SYRMA-3D project.

5.2   The Effect of Pixel Size

The other key parameter in determining the effectiveness of propagation-based imaging and phase-retrieval filtration is the detector pixel size. Intuitively this can be explained by considering that phase effects in PB imaging emphasize the sample high spatial frequencies, therefore requiring for a high spatial resolution detector. Moreover, an effective detection of edge-enhancement effects, arising upon propagation, determines the effectiveness of the subsequent phase-retrieval algorithm in producing a high SNR image without introducing an excessive smoothing. Hereinafter the notion of detector featuring a 'high spatial resolution' will be identified with 'small pixel size'. This simplification is not rigorously valid for indirect-conversion detectors, whereas it is fairly accurate for many direct-conversion photon-counting detectors as Pixirad-8, where the PSF width is mainly determined by the pixel size. By making use of the noise model previously introduced, in this section the effect of pixel size on image noise will be studied, and theoretical results will be compared with experimental data. Some of the results hereby presented have been documented in [10].

5.2.1   Noise Dependence on Pixel Size in Propagation-Based CT

Regardless of the imaging modality (attenuation, propagation-based etc.) noise magnitude CT images is strongly dependent on the detector pixel size. Starting from the model described by Eq. (5.5) and isolating only the terms related to the pixel size, the variance measured in reconstructed tomographic image reads

$$\sigma^2 \propto \frac{f(A; d/h)}{h^4} \tag{5.13}$$

where the numerator is function of the pixel size only through the parameter A, as described by Eq. (5.7). When no PhR is applied (i.e. $A = 0$), as in case of conventional attenuation-based CT, the previous equation implies that image noise increases with the inverse of the square of the pixel size [4]. Given the steep dependence between image noise and pixel size, high-resolution CT images with acceptable noise levels cannot be obtained when constraints in terms of radiation dose or scan time are present, as in clinical or animal studies. Conversely, when PhR is applied (i.e. $A \neq 0$), noise dependence on the pixel size is much shallower, being mitigated by the function f, as shown in Sect. 5.1.1.2. In particular, f is monotonically decreasing for increasing values of A, hence, being $A \propto 1/h^2$, for decreasing pixel sizes.

Assuming, as done in the previous section, a bi-dimensional sinc function MTF, Shepp-Logan reconstruction filter and linear interpolation during the backprojection process, the CT image noise can be computed as a function of pixel size by making use of Eqs. (5.5) and (5.6). The numerical results, spanning a pixel size interval from 10 μm to 1000 μm and the same propagation distances reported in the previous section, are shown in Fig. 5.6: interestingly, for all propagation distances, the difference in noise between images reconstructed with or without PhR is amplified at smaller pixel sizes, meaning that the noise-reduction effect due to PhR becomes more effective as the pixel size decreases. On the other hand, at large pixel sizes, the noise level of PhR images asymptotically converge to the non-PhR case, thus the application of PhR does not entail any improvement in terms of SNR. This can be easily understood as almost no (high-frequency) phase effects arising during the propagation process can be detected if the pixel size is too large. In addition, it is worth noting that the differences in the trends of the two curves are further exacerbated by the propagation distance, coherently with the results presented in Sect. 5.1.3.

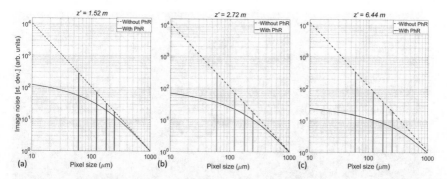

Fig. 5.6 Tomographic image noise as a function of the pixel size computed at propagation distances of 1.52 m (**a**), 2.72 m (**b**) and 6.44 m (**c**). Dashed lines refer to images without PhR, solid lines to images with PhR. For each plot the image noise has been normalized to the PhR case at pixel size of 1000 μm. Vertical lines indicate the pixel sizes of experimental data, whose results are reported in Fig. 5.8

In analogy with the limitations on the SNR increase with increasing propagation distances, the noise cannot be indefinitely decreased by having arbitrarily small pixel sizes since the presented mathematical formulation holds in the large Fresnel number approximation (N_F decreases with the square of the pixel size).

5.2.2   Noise Dependence on Pixel Size: Experimental Results

To test the effect of pixel size on experimental data, the same breast specimen presented in the previous section, scanned at three propagation distances, has been used. In order to achieve different pixel sizes projection images have been re-binned by factors of 1, 2, 3, and 4 prior to PhR, resulting in pixel pitches of 60 (native spacing), 120, 180 and 240 μm. Following the re-binning procedure, projections are processed according to the reconstruction pipeline described in Sect. 3.7, and the SNR is measured within a homogeneous glandular detail for both phase-retrieved and non-phase-retrieved datasets.

Figure 5.7 shows a detail of the reconstructed volume at different pixel pitches, both without (top row) and with (bottom row) PhR. From the images it is clear that the noise reduction due to PhR is crucial to drastically reduce image noise, thus improving detail visibility, at a pixel size of 60 μm, while its effect is less and less noticeable for larger pixel sizes. In particular, at 240 μm pixel size, phase retrieval does not produce a relevant gain in SNR and its application can be avoided without impairing the visibility of glandular structures.

From the reconstructed datasets the SNR gain factor has been computed, following the definition of Eq. (5.10), and compared with the numerical results obtained from the plots in Fig. 5.6. The comparison between the model predictions and the experimental data is reported in Fig. 5.8. From the plot an excellent agreement between

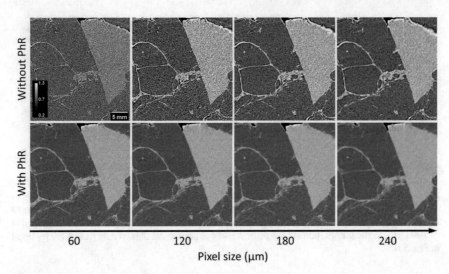

Fig. 5.7 Crop of a reconstructed tomographic slice showing a fibroglandular detail embedded in adipose background acquired at 6.44 m of propagation distance. Each column refers to a given pixel size while each row specifies whether the PhR is applied or not. All the images are windowed on the same gray level scale (inset of the top-left image) to facilitate the comparison. Reproduced from [10] by permission of IOP Publishing

model and data is observed for all propagation distances. Of note, the experimental points obtained with the 60 μm pixel pitch correspond to the ones in Fig. 5.4, where the dependence on propagation distance was studied. It is interesting to observe that the increase in the pixel size is associated to a major decrease of the SNR gain due to PhR. This trend is more pronounced for larger propagation distances: specifically, at 6.44 m the gain increases by a factor of 7 going from the largest to the smallest pixel pitch, while at 1.52 m its increment is in the order of a factor 3.5.

In light of the results presented in this section, which are supported by a rigorous mathematical model, it is clear that going towards small pixel sizes is crucial to fully exploit the noise reduction capabilities of phase retrieval. Going back to the problem of achieving low-dose tomographic images with a high spatial resolution, the use of propagation-based imaging coupled with a small pixel size detector can be an invaluable tool to overcome, or at least mitigate, visibility issues related to excessive image noise. In this context, the development of high-efficiency photon counting detector with smaller pixels, coupled with suitable on-chip processing strategies to compensate for charge sharing effects [11, 12], will be of great importance for a wider and more efficient use of PB imaging in biomedical applications.

Fig. 5.8 SNR gain as
function of pixel size derived
from experimental data
(points) and from the
theoretical model (lines) at
three propagation distances

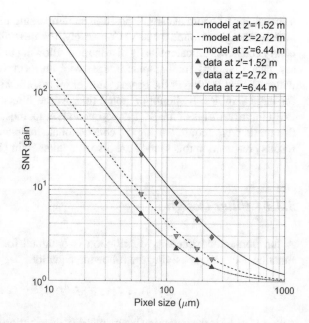

5.3 Beam Profile Optimization: Flattening Filter

In addition to the high coherence, X-rays produced by synchrotrons are generally
several orders of magnitude more intense with respect to conventional sources. For
this reason, many bio-medical imaging applications, as BCT, require beam filtration
to deliver acceptable dose levels [13, 14]. As described in Sect. 3.6, this is typically
performed by inserting aluminum sheets or slabs that reduce the overall beam intensity without affecting its spatial distribution (or 'shape'). Specifically, the vertical
(i.e. orthogonal to the electrons' orbit plane) intensity profile of a synchrotron beam
produced by bending magnets can be described by a Gaussian distribution [15]. This
leads to an undesired non-uniform dose distribution on the sample in the vertical
direction. In terms of image quality, this translates into a non constant SNR, which
decreases moving from the central maximum of the beam towards the tails. To limit
such non-uniformity, in many experiments only the central part of the beam is used
for imaging purposes, while the tails are filtered out by absorbing (e.g., made of
tungsten) slits. Despite being easy to implement, this approach is not optimal in
sight of any application, especially in-vivo, requiring the scan of large samples as
the reduction of the vertical beam dimension entails an increase in the number of
vertical scans required to image a large volume and, as a consequence, an increase
in the overall scan duration.

To overcome the non-uniformity, while using the full beam vertical dimension,
an *ad-hoc* parabolic shaped flattening filter has been designed and implemented.
Up to now, a slit system made of Densimet® (tungsten alloy), coupled with planar aluminum filters, has been routinely used. This system defines a vertical beam

dimension of 3.5 mm at sample position encompassing intensity variations of about 30% at energies around 30 keV. Conversely, the new filtration system produces a nearly constant vertical intensity distribution, allowing uniform radiation dose delivery, hence yielding tomographic images with uniform SNR, as well as to use of a wider vertical portion of the beam (5 mm or more), allowing for scan time reduction for large samples. It should be noted that, as the filter development is one of the latest improvements of the BCT experimental setup, most of the images presented in this thesis were acquired using the conventional slits/planar filtration system. Many results presented in this section are also documented in [16].

5.3.1 Filter Design

A flat transmitted intensity distribution is obtained for a filter, described by the function $F(y; E)$, satisfying the following equation:

$$I_f(y) = I(y; E)e^{-\mu_f(E)F(y;E)} = k \tag{5.14}$$

where $I(y, E)$ is the incoming beam intensity distribution along the vertical direction y, $\mu_f(E)$ is the energy (E) dependent attenuation coefficient of the filter and $I_f(y)$ is the flattened transmitted beam, whose intensity is equal to a transmitted fraction k of the maximum of the input beam. By assuming that the unfiltered beam has a Gaussian vertical spatial distribution, $I(y; E) \propto \exp\left(-\frac{y^2}{2\sigma_y^2(E)}\right)$, with an energy dependent standard deviation $\sigma_y(E)$, the filter shape can be computed by solving Eq. (5.14), and it reads

$$F(y; E) = -\frac{y^2}{2\sigma_y^2(E)\mu_f(E)} - \frac{\ln k}{\mu_f(E)} \tag{5.15}$$

Therefore, the desired filter has a parabolic shape whose depth (d_f, i.e. size along the beam propagation direction) and height (h_f, i.e. size along the vertical dimension of the beam) are, respectively

$$d_f = \frac{|\ln k|}{\mu_f(E)}, \quad h_f = 2\sigma_y(E)\sqrt{2|\ln k|} \tag{5.16}$$

At this point it can be noted that the filter depth depends both on the filter material, through its attenuation coefficient, and on the desired intensity fraction of the impinging beam. Conversely, the filter height is dependent on the beam's vertical dimension and its intensity fraction while it is independent of the filter material.

The implemented filter is made of aluminum and it has been designed for an energy of 30 keV. The beam standard deviation at sample position, i.e. 30 m from the X-ray source, corresponding to 30 keV is $\sigma_y(30 \text{ keV}) = 1.8$ mm. Since the filter is

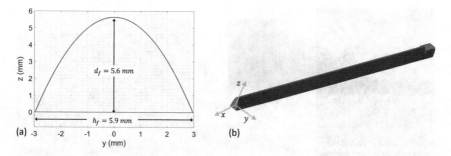

Fig. 5.9 Filter parabolic profile optimized for 30 keV beam and 26 m from the source at SYRMEP beamline (**a**) and its CAD design (**b**)

positioned 4 m upstream with respect to the sample, a standard deviation of 1.6 mm has been considered to compensate for the beam magnification. The transmission factor is chosen to be $k = 18\,\%$, providing sufficient flux for delivering (mean glandular) dose rates up to 0.5 mGy/s to large ($\sim$10 cm) breast samples. Considering the standard scan time (40 s) for acquiring BCT images, this results in doses up to 20 mGy, which corresponds to the 'high-image quality' (dedicated to surgical specimens) modality of the SYRMA-3D protocol [17]. For this reason, extra filtration composed of aluminum sheets is needed to match the clinical target dose level of 5 mGy for in-vivo applications. Given k and σ parameters, the filter has been modelled via a computer-aided design (CAD) software and manufactured with a computer numerical control machine (see Fig. 5.9).

As a general remark, the energy dependence of the filter shape can be viewed as a practical drawback since, in principle, each energy would require a dedicated design. As it will be clarified in the next section, the proposed filter is proven to be sufficiently flexible for energies around 30 keV. In fact, the filter yields a beam which is more homogeneous with respect to the standard planar filtration system in a range of energies between 28 and 32 keV, which is of interest for the breast CT application. If a similar degree of flexibility is desired at lower energies, the same filtration approach could be adapted by using lighter filtering materials, whose attenuation coefficient is Compton dominated. As an example, plastic filters (i.e. $6 < Z_{eff} < 7$) would offer some energy flexibility down to about 20 keV.

5.3.2 Filter Tests in Planar and Tomographic Configurations

The flattening filter has been tested at 3 different energies of 28, 30, and 32 keV. As shown in Fig. 5.10, when used at the design energy of 30 keV (panels (b) and (e)), the filter ensures a beam profile with intensity fluctuations up to 5% and a height of 5.5 mm, whereas the unfiltered beam (a) has, in the same spatial range, a maximum intensity variation of more than 60% around the mean value. Moreover, even

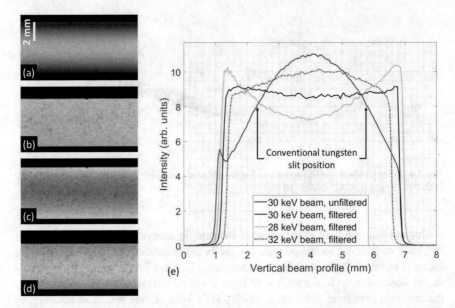

Fig. 5.10 Images of the beam with no filtration (**a**) and flattening filter at 30 keV (**b**), 28 keV (**c**) and 32 keV (**d**). Vertical profiles (**e**) of the beams reported in panels (**a**)–(**d**). Black arrows the position corresponding to the tungsten slits system used for the clinically-oriented imaging acquisitions so far. Profiles in (**e**) are normalized to their area

considering only the portion of the beam (3.5 mm) that would have been transmitted by the slit system, the intensity variation in the unfiltered beam is still around 30%. When employed at a beam energy of 28 keV (c), the filter introduce an excessive attenuation in the central part, yielding a cup-shaped profile. Anyway, the observed intensity variation over the entire beam height (5.7 mm) is of the order of 30%, that is half of the variation of the unfiltered beam in the same spatial range. The opposite behaviour is found for the 32 keV irradiation (d): in this case the maximum intensity fluctuation across the whole beam height (5 mm) is of the order of 15%, about 4 times smaller if compared to the unfiltered beam.

To demonstrate the effectiveness of the filter in a realistic scenario, the tomographic reconstructions of two mastectomy samples with similar sizes, imaged with and without using the flattening filter, have been compared. Coherently with all the scans presented in the next chapter, 5 mGy of mean glandular dose were delivered to both samples while the selected scan energy was 32 keV (which is not the optimal energy for the described filter). The results are summarized in Fig. 5.11 showing, in panels (a), (b), the reconstructed details of the sample acquired with no flattening filter, considering slices corresponding to the central and the tail regions of the beam, respectively. In the same way, panels (c), (d) show slices of the sample acquired with the flattening filter at the center and at the edge positions of the beam. For both samples, the SNR, defined as the ratio between the mean and the standard deviation of gray values within a selected region of interest, is measured within a glandular

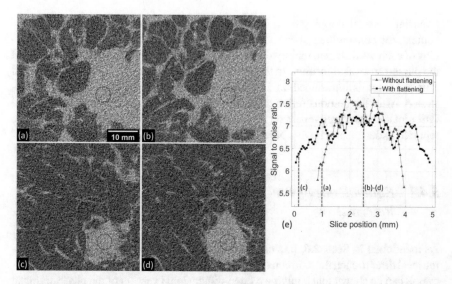

Fig. 5.11 Details of reconstructed slices corresponding to the central portion and to the edge of the vertical field-of-view obtained with conventional planar filtration system ((**a**) edge, (**b**) center) and with the flattening filter ((**c**) edge, (**d**) center), respectively. Plot of SNR as a function of slice position measured within the dashed circles in (**a**)–(**d**) for both filtering configurations. Light-blue dashed lines represent slice positions of panels (**a**)–(**d**)

detail as a function of the slice position, as reported in panel (e). As the plot shows, the use of the flatting filter brings to a smoother dependence of the SNR on the slice position compared to the conventional planar filtration. Because the flattening filter allows for a more even distribution of the radiation dose, the measured SNR is lower in the center and higher at the edges of the beam with respect to the conventional filtration case. Moreover, the wider portion of the beam useful for imaging (from slightly more than 3 mm to 5 mm for the case reported in Fig. 5.11e) translates into a reduction of the scan time for imaging the whole volume of approximately 40%. This is of great importance in view of the clinical implementation, as it will limit patient discomfort, thus motion-related artifacts, and it will improve the examination throughput.

5.4 Post-reconstruction Phase Retrieval in Multi-stage Scans

As described in the previous section, the limited vertical dimension of synchrotron X-ray beams usually requires multiple vertical steps (or stages) to image large samples, as in the BCT case. Moreover, photon-counting detectors are usually composed by mono-dimensional arrays of individual sensors few centimeters in height, thus requiring vertical stepping even in case of an arbitrarily wide X-ray beam. In this context, the conventional reconstruction pipeline can introduce artifacts at the margins of each vertical step mainly due to boundary effects arising in the application of the phase-retrieval algorithm. In this section a post reconstruction three-dimensional PhR approach is introduced, and its ability to cope with these artifacts is demonstrated. After the demonstration of its theoretical equivalence with the conventional PhR pipeline, its effectiveness on experimental images is demonstrated. Some of the reported experimental results have been published in [18].

5.4.1 Equivalence of Pre- and Post-reconstruction Phase Retrieval

As mentioned in Sect. 2.6, it is common practice to apply a bi-dimensional phase-retrieval filter to each flat-corrected projection prior to the actual reconstruction. However, it can be shown that applying a three-dimensional version of the phase-retrieval filter after tomographic reconstruction leads to theoretically equivalent results. Intuitively, this can be understood as both PhR and tomographic reconstruction are linear (and commutative) filters in the Fourier space. A rigorous formal demonstration of this has been given by Ruhlandt and Salditt [19] under the 'weak object' approximation, which assumes both attenuation and phase-shift terms in the complex transmission function to be small (see Eq. (2.4)).

Actually, when dealing with near-field PB imaging and Paganin's PhR algorithm, the weak attenuation assumption can be dropped and the mathematical formulation of the three-dimensional PhR filter can be derived *mutatis mutandis* from the bi-dimensional case. In fact, in Sect. 2.6 it was shown that, under the weak phase contrast hypothesis, the tomographic map obtained from PB (i.e. with no PhR) projections can be written as

$$o^{PB}(x, y, z) = \mu(x, y, z) - z_1 \nabla^2_{xyz} \delta(x, y, z) \tag{5.17}$$

At this point the homogeneous object condition (i.e. δ/β is a known constant parameter) can be inserted and, by conveniently re-writing $\delta = \delta\mu/(2k\beta)$, the previous equation becomes

$$o^{PB}(x, y, z) = \left[1 - \frac{z_1\delta}{2k\beta} \nabla^2_{xyz}\right] \mu(x, y, z) \tag{5.18}$$

where the term enclosed in square brackets is immediately identified with the three dimensional version of the froward propagation operator (H) defined in Eq. (2.23). In other words, Eq. (5.18) mathematically describes the propagation of the entire three-dimensional object in the near field which is equivalent to the propagation of each individual bi-dimensional projection. Following this analogy, the inverse operator, that is the phase retrieval, will simply be the three-dimensional extension of the expression reported in Eq. (2.24):

Fig. 5.12 Collage of a sample projection for each of the ten considered vertical stages. Due to the limited vertical size of the beam, the height of each projection consists in 51 pixels. The projections cannot be easily stitched together to compose a single projection of 510 pixels height because of the unknown angular shift and actual angular range covered induced by the continuous mode acquisition: the registration is performed within the reconstruction step

$$\tilde{H}_{3D} = \left[1 + \frac{z_1 \delta}{2k\beta}(v_1^2 + v_2^2 + v_3^2)\right]^{-1} \tag{5.19}$$

where (v_1, v_2, v_3) are the Cartesian coordinates in the three-dimensional Fourier space.

5.4.2 Bi- Versus Three-Dimensional Phase Retrieval Pipelines

When a large volume is scanned with multiple vertical stages, suitable strategies are required for the inherent issue of image stitching [20, 21] in order to correctly create the reconstructed volume of the whole object. Projection stitching typically requires the determination of the center of rotation and in practical multi-stage tomography it might slightly vary from one vertical stage to another. Moreover, when considering continuous acquisition mode as in the case of BCT, the determination of the exact angular range covered by the scan is needed for the stitching procedure and, in general, it is different for each stage, as reported, for instance, in Fig. 5.12. Both these issues are usually tackled by registering and stitching the reconstructed slices rather than operating on the projections. As aforementioned, such a procedure implies that the PhR is applied independently to each projection of each vertical stage, generally introducing periodic artifacts in the lateral views of the tomographic volume, in correspondence with the junction slices between two adjacent stages. The reason for those artifacts lies in the absence of knowledge about the neighboring pixels of the upper and lower part of each projection image when applying the bi-dimensional (2D) independent stage-by-stage processing. In facts, the 2D PhR approach cannot consider the real information coming from the adjacent vertical stages. In most cases this shortcoming is partly overcome by replicate padding of each projection

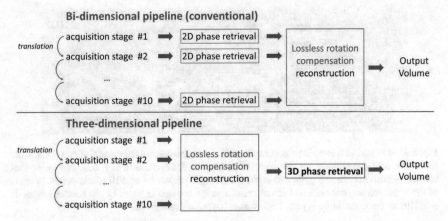

Fig. 5.13 Sketch of the two computational pipelines compared in this section. The term 'lossless rotation compensation' refers to the angular offset to be specified as additional input to the reconstruction algorithm. In this way image registration is performed by 'rolling' the sinogram to match a desired rotation angle without requiring any image interpolation

which mitigates for the absence of this information. On the other hand, the post-reconstruction three-dimensional (3D) PhR pipeline allows to perform the phase retrieval on the whole volume after the registration of each reconstructed vertical stage, thus inherently solving the missing information issue. A sketch of both 2D and 3D PhR pipelines is shown in Fig. 5.13.

5.4.3 Quantitative Comparison on a Large Breast Specimen

To demonstrate the effectiveness of the post-reconstruction 3D PhR, the two processing pipelines have been compared, based on images of a large mastectomy specimen. The sample, featuring a diameter of 9 cm and a height of 3 cm, contained an infiltrating ductal carcinoma with a diameter of about 1.2 cm. After positioning the sample in the patient support (1.6 m of propagation distance), it was scanned at 38 keV with 10 vertical steps. Each of the 1200 angular projections acquired for every vertical position was cropped to a dimension 2150×51 pixels, resulting, after the stitching procedure, in a final volume of $2150 \times 2150 \times 510$ voxels.

The set of projections was processed following either the 2D PhR or the 3D PhR pipelines (two-materials PhR, $(\delta_1 - \delta_2)/(\beta_1 - \beta_2) = 1083$). The two reconstructed volumes have been quantitatively compared in terms of spatial resolution, contrast (C) and contrast-to-noise ratio (CNR) by considering the central slice of a given 51 slices stack. Spatial resolution was measured, as described in Sect. 5.1.2, starting from the three intensity profiles reported in blue in Figs. 5.14 and 5.15. The two circular regions (one within glandular tissue referred to as A and the other one within adipose tissue referred to as B) reported in the same figures were used to compute the mean

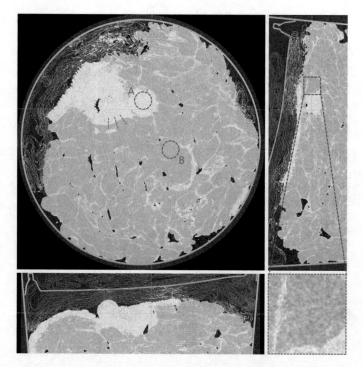

Fig. 5.14 Orthogonal views of the output volume when the pipeline with 2D phase retrieval is applied. A close-up (bottom-right) is reported to highlight the observed artifact at the interfaces between adjacent vertical stages. Colored segments or ROIs were used for quantitative analysis as described in text

$\langle I \rangle$ and the standard deviation σ of the gray levels. From these quantities the contrast was determined as reported in Eq. (5.12), while the CNR was computed as

$$\text{CNR} = \frac{\langle I_A \rangle - \langle I_B \rangle}{\sigma_b} \tag{5.20}$$

In addition to these metrics, with the aim of highlighting the artifact at the interface of adjacent reconstructed stages, the standard deviation measured from a line ROI covering 121 voxels (green lines in Figs. 5.14 and 5.15) was evaluated for each reconstructed slice and plotted against the corresponding vertical position.

Figures 5.14 and 5.15 show the transverse, i.e. orthogonal to the rotation axis, and lateral views of the entire reconstructed volume for the 2D and 3D PhR cases, respectively. When comparing the transverse slices, which correspond to the center of a vertical stage, almost no differences between the two pipelines are observed. This qualitative evaluation is confirmed by the numerical results of the quantitative analysis, summarized in Table 5.3. The analysis revealed almost identical spatial resolution, contrast and CNR for both the considered cases. For the sake of com-

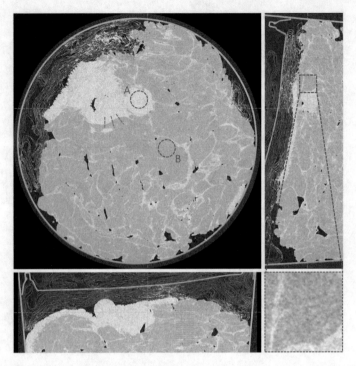

Fig. 5.15 Same as Fig. 5.14 but reconstructed through the 3D phase retrieval. The close-up shows the lack of artifacts between adjacent vertical stages

Table 5.3 Quantitative comparison between 2D and 3D phase-retrieval pipelines

	Spatial resolution FWHM (mm)	Contrast C (%)	Contrast-to-noise ratio CNR
2D pipeline	0.117 ± 0.025	34.4	3.46
3D pipeline	0.118 ± 0.025	34.3	3.46

pleteness the line profiles along with the fit functions used to estimate the spatial resolution are reported in Fig. 5.16: of note is that, considering an effective pixel size of 57 μm, the values of FWHM found in this analysis well compares with the ones reported in Sect. 5.1.3.

While the two approaches yield substantially identical results when considering transverse slices far from the margins of a vertical stage, as anticipated from the formal equivalence between 2D and 3D PhRs, major differences are found in the junction slices across two stitched vertical stages. This can be observed in the close-up images of Figs. 5.14 and 5.15 (bottom-right panels) and, quantitatively, from the plot reported in Fig. 5.17, where a periodic spike in the measured standard deviation can be clearly noticed every 51 slices in the case of the pre-reconstruction 2D PhR approach, while no artifacts are visible in the post-reconstruction 3D PhR case. Of note, for all the transitions across different stages, the artifact involves 3 or 4 slices,

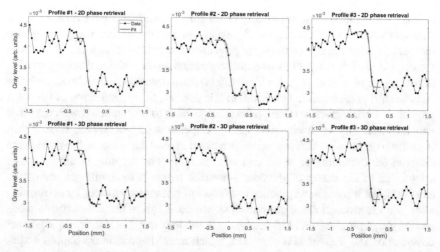

Fig. 5.16 Plots of the fit used for the assessment of the spatial resolution on a reconstructed slice for the 2D case (top row) and 3D case (bottom row). The three considered profiles are highlighted in blue in Figs. 5.14 and 5.15

Fig. 5.17 Plot of the standard deviation of the gray levels with reference to the green lines in Figs. 5.14 and 5.15. A spike every 51 slices is noticeable for the 2D phase retrieval

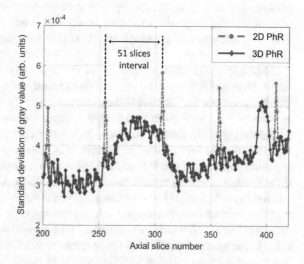

meaning that, given the displacement dimension of the considered vertical stages, about 6 to 8% of the volume presents an undesired increase of noise due to the bidimensional pipeline. As previously mentioned, the advantage of 3D PhR over the 2D approach is the possibility of using the whole object information by filtering the reconstructed volume rather then relying on a single, vertically limited projection.

It should be stressed that, if a full-height (510 pixels) projection could be composed by tiling each vertical stage, this artifact would result similarly compensated through the conventional 2D PhR approach. However, several factors hamper this stitching process. Firstly, alignment issues due to imperfect relative positioning of detector and rotating stage are usually better compensated during the reconstruction step with inspection of the computed images. As a matter of fact, recognizing geometrical misalignment from the input projections only is a difficult task. Moreover, the combination of continuous rotation acquisition mode with the limited precision of rotating stage motors (e.g., backlash effect) often determines an unknown relative angular shift at each vertical position. In this context, the identification of the exact projection from a given vertical position acquisition to be combined with the other vertical stages requires horizontal flipping. Depending on whether the angular shift is positive or negative, this procedure is needed at least for some of the projections either at the beginning or at the end of each scan. The flipping, in turn, requires knowledge of the center of rotation: automatic methods for its the determination exist [22], but it is common practice to assess its correctness visually, by performing a few test reconstructions of just one slice. Moreover, given that automatic methods require projections at 0 and 180 degrees as input, they may fail if the complete/exact coverage of 180 degrees is not granted, which might happen in continuous acquisitions (e.g., in BCT case). Alternatively, image correlation techniques can be used for automatic angular shift assessment, but the correctness of their output is usually better supervised by an expert user checking a reconstructed slice rather than projections. For these reasons, the application of the conventional 2D phase-retrieval to a set of stitched projections requires in any case some preliminary reconstruction. The 3D phase-retrieval approach allows to skip the stitching phase and the related challenges.

Although an accurate comparison of the two approaches in terms of computational efficiency is beyond the scope of this section, some qualitative evaluations can still be made. As the 2D PhR is, in principle, an on-line process, it can be applied as soon as each projection is collected without waiting for the acquisition of the whole tomographic dataset, resulting in a faster experimental pipeline. However, this procedure is seldom used in experimental practice. On the other hand, the 3D approach requires the whole reconstructed volume as input, which means waiting for the collection of all the projections (off-line procedure). Additionally, since the whole volume has to be loaded for processing, 3D phase retrieval requires a large amount of memory. For instance, the stacked volume considered in this section is composed of $2150 \times 2150 \times 510$ voxels, which means a 32-bit floating point matrix of about 8.8 GB. Furthermore, given that 3D phase retrieval relies on three-dimensional Fourier filtering, 3D signal padding is fundamental to avoid cross-talk between opposite sides of the volume. Compared with the 2D case, where for each projection only horizontal and vertical padding is performed, the 3D padding accounts also for an additional dimension, leading to a larger number of matrix elements to be processed. For this reason, post-reconstruction PhR can be generally considered more computationally demanding if compared to the conventional approach, therefore its use must be evaluated depending on available computational power and dataset dimension.

References

1. Nesterets YI, Gureyev TE (2014) Noise propagation in X-ray phase-contrast imaging and computed tomography. J Phys D: Appl Phys 47(10):105402. https://doi.org/10.1088/0022-3727/47/10/105402

2. Gureyev TE, Nesterets YI, Kozlov A, Paganin DM, Quiney HM (2017) On the "unreasonable" effectiveness of transport of intensity imaging and optical deconvolution. JOSA A 34(12):2251–2260. https://doi.org/10.1364/JOSAA.34.002251

3. Nesterets YI, Gureyev TE, Dimmock MR (2018) Optimisation of a propagation-based X-ray phase-contrast micro-CT system. J Phys D: Appl Phys 51(11):115402. https://doi.org/10.1088/1361-6560/aa5d3d

4. Davis GR (1994) The effeCT of linear interpolation of the filtered projections on image noise in X-ray computed tomography. J X-ray Sci Technol 4(3):191–199. https://doi.org/10.3233/XST-1993-4303

5. Kitchen MJ, Buckley GA, Gureyev TE, Wallace MJ, Andres-Thio N, Uesugi K, Yagi N, Hooper SB (2017) CT dose reduction factors in the thousands using X-ray phase contrast. Scient Reports 7(1):15953. https://doi.org/10.1038/s41598-017-16264-x

6. Gureyev TE, Nesterets YI, Stevenson AW, Miller PR, Pogany A, Wilkins SW (2008) Some simple rules for contrast, signal-to-noise and resolution in in-line X-ray phase-contrast imaging. Opt Exp 16(5):3223–3241. https://doi.org/10.1364/OE.16.003223

7. SYRMEP. SYRMEP specifications (2016). www.elettra.trieste.it/lightsources/elettra/elettra-beamlines/syrmep/specification.html

8. Taylor JA (2018) TS imaging. http://ts-imaging.science.unimelb.edu.au/Services/Simple/

9. Brombal L, Donato S, Dreossi D, Arfelli F, Bonazza D, Contillo A, Delogu P, Di Trapani V, Golosio B, Mettivier G et al (2018) Phase-contrast breast CT: the effect of propagation distance. Phys Medi Biol 63(24): 24NT03. https://doi.org/10.1088/1361-6560/aaf2e1

10. Brombal L (2020) Effectiveness of x-ray phase-contrast tomography: effects of pixel size and magnification on image noise. J Instrum 15(01):C01005. https://doi.org/10.1088/1748-0221/15/01/C01005

11. Di Trapani V, Bravin A, Brun F, Dreossi D, Longo R, Mittone A, Rigon L, Delogu P (2018) Characterization of noise and efficiency of the pixirad-1/pixie-iii cdte x-ray imaging detector. J Instrum 13(12):C12008. https://doi.org/10.1088/1748-0221/13/12/C12008

12. Gimenez E, Ballabriga R, Campbell M, Horswell I, Llopart X, Marchal J, Sawhney K, Tartoni N, Turecek D (2011) Study of charge-sharing in medipix3 using a micro-focused synchrotron beam. J Instrum 6(01):C01031. https://doi.org/10.1088/1748-0221/6/01/C01031

13. Bravin A, Coan P, Suortti P (2012) X-ray phase-contrast imaging: from pre-clinical applications towards clinics. Phys Med Biol 58(1):R1. https://doi.org/10.1088/0031-9155/58/1/R1

14. Rigon L (2014) X-ray imaging with coherent sources. In: Brahme A (ed) Comprehens Biomed Phys 2:193–216. Elsevier. https://doi.org/10.1016/B978-0-444-53632-7.00209-4

15. Viccaro PJ (1991) Power distribution from insertion device X-ray sources. In: Advanced X-Ray/EUV radiation sources and applications, vol 1345, pp 28–38. International Society for Optics and Photonics. https://doi.org/10.1117/12.23298

16. Donato S, Arfelli F, Brombal L, Longo R, Pinto A, Rigon L, Dreossi D (2020) Flattening filter for gaussian-shaped monochromatic x-ray beams: an application to breast computed tomography. J Synchrotron Radiat in press. https://doi.org/10.1107/S1600577519005502

17. Piai A, Contillo A, Arfelli F, Bonazza D, Brombal L, Cova MA, Delogu P, Trapani VD, Donato S, Golosio B, Mettivier G, Oliva P, Rigon L, Taibi A, Tonutti M, Tromba G, Zanconati F, Longo R (2019) Quantitative characterization of breast tissues with dedicated CT imaging. Phys Med Biol 64(15):155011. https://doi.org/10.1088/1361-6560/ab2c29 Aug

18. Brun F, Brombal L, Di Trapani V, Delogu P, Donato S, Dreossi D, Rigon L, Longo R (2019) Post-reconstruction 3D single-distance phase retrieval for multi-stage phase-contrast tomography with photon-counting detectors. J Synchrotron Radiat 26(2). https://doi.org/10.1107/S1600577519000237

19. Ruhlandt A, Salditt T (2016) Three-dimensional propagation in near-field tomographic X-ray phase retrieval. Acta Crystallographica Sect A: Foundat Adv 72(2):215–221. https://doi.org/10.1107/S2053273315022469

20. Kyrieleis A, Ibison M, Titarenko V, Withers P (2009) Image stitching strategies for tomographic imaging of large objects at high resolution at synchrotron sources. Nucl Instrum Methods Phys Res Sect A: Acceler Spectromet Detectand Assoc Equip 607(3):677–684. https://doi.org/10.1016/j.nima.2009.06.030

21. Vescovi R, Du M, Andrade VD, Scullin W, Gürsoy D, Jacobsen C (2018) Tomosaic: efficient acquisition and reconstruction of teravoxel tomography data using limited-size synchrotron X-ray beams. J Synchrotron Radiat 25(5). https://doi.org/10.1107/S1600577518010093

22. Vo NT, Drakopoulos M, Atwood RC, Reinhard C (2014) Reliable method for calculating the center of rotation in parallel-beam tomography. Opt Express 22(16):19078–19086. https://doi.org/10.1364/OE.22.019078

Chapter 6
Three-Dimensional Imaging: A Clinically Oriented Focus

Taking advantage of the optimizations and procedures introduced in the previous two chapters, this chapter demonstrates the imaging results presently achievable at the SYRMEP beamline, with a focus closely oriented to the clinical application of propagation-based BCT.

Of note, most of the PBBCT data documented in literature to date have been limited to breast specimens featuring a small thickness [1–3]. Nonetheless, in order to demonstrate the advantages over conventional imaging, fully three-dimensional CT datasets must be produced by imaging the whole volume as done, for instance, with other phase-sensitive techniques [4–6]. In recent publications, by both the Italian and Australian collaborations [7, 8], the first full 3D reconstructions of breast specimens imaged using PB technique at clinically acceptable dose levels have been shown. In the following, based on full volume scans of three large mastectomy/lumpectomy samples, several features of PBBCT images, as 3D visualization and convenient data processing, are presented. In addition, to investigate the foreseeable diagnostic benefits associated with PBBCT, images are compared with the currently available standard clinical techniques: as a matter of fact, breast-cancer detection relies mostly on (planar) mammographic images while the intra-operative or post-surgery analysis of the resected tissue is performed by means of histological examination. In this context, two cases are compared with conventional X-ray mammography imaging and, in one case, the matching between histological and low dose PBBCT images is demonstrated. Some contents of this chapter are based on the results published in [9].

6.1 Samples and Acquisition Parameters

The work reported in this chapter was carried out following the Directive 2004/23/EC of the European Parliament and of the Council of 31 March 2004 on setting standards of quality and safety for the donation, procurement, testing, processing, preservation,

© The Editor(s) (if applicable) and The Author(s), under exclusive license
to Springer Nature Switzerland AG 2020
L. Brombal, *X-Ray Phase-Contrast Tomography*, Springer Theses,
https://doi.org/10.1007/978-3-030-60433-2_6

storage and distribution of human tissues. The presented images were acquired as to guide the pathologist in the lesion localization during histological preparation, according to the standard procedures of the clinic operative unit (U.C.O.) of the Anatomy and Histology Department of the University Hospital of Cattinara, Trieste. The samples were prepared from specimens of breast mastectomy and lumpectomy sent to the clinic operative unit, where they were sealed in a vacuum bag after formalin fixation. Within the energy range of interest, this process is expected not to produce substantial alterations in contrast between adipose and fibroglandular/tumoral tissue, as reported in literature [10]. Three surgical specimens containing cancer were analyzed and described by expert pathologists as follows:

- sample A is a left simple mastectomy from a 86 year old woman. The histological exam revealed a high-grade infiltrating solid carcinoma with a maximum diameter of 8 cm;
- sample B is a lumpectomy in left upper inner breast from a 84 year old woman. The histological exam revealed a moderate-grade infiltrating ductal carcinoma with a maximum diameter of 2.4 cm with a central sclerotic area;
- sample C is a right simple mastectomy from a 77 year old woman. The histological exam revealed a moderate-grade infiltrating ductal carcinoma with a maximum diameter of 9 cm.

As described in Sect. 3.5, the samples were imaged by acquiring 1200 projections in continuous rotating mode over 180°, at the maximum detector frame rate of 30 Hz. Scans were performed in 40 s, corresponding to an angular speed of 4.5 degrees/second. Due to the small vertical beam dimension (3.5 mm, FWHM), to acquire the full volume many scans (8–14) at different vertical sample positions were collected, resulting to a total scan time ranging from 5 to 9 min. By adjusting the beam intensity through planar aluminum filters (see Sect. 5.3), 5 mGy of mean glandular dose were delivered, and the specimens were imaged at 32 keV. To facilitate the comparison with results of other groups which use different dosimetric protocols [3, 8], in the following the entrance air kerma is declared for each image. Prior to reconstruction the single-material PhR was applied with $\delta/\beta = 2308$, corresponding to breast equivalent tissue.

6.2   3D BCT Reconstructions and Comparison Conventional Imaging

6.2.1   Sample A

The reconstructed three-dimensional volume of the sample A is reported in Fig. 6.1, where the three orthogonal view planes, i.e. coronal, sagittal and transverse (see inset) are displayed. In order to maintain the conventional anatomical planes, the one orthogonal to the rotation axis, usually referred to as transverse, is here identified as

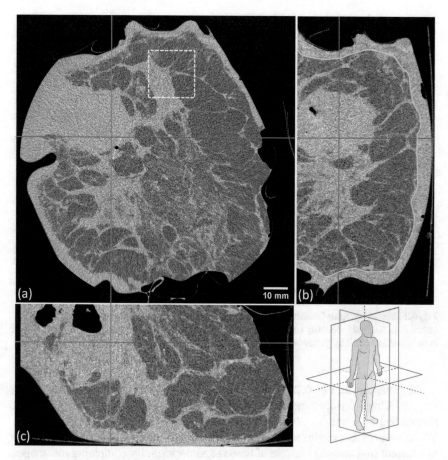

Fig. 6.1 Coronal (**a**), sagittal (**b**), transverse (**c**) views of the sample A. Line markers are centered in the bulk of the biggest tumoral focus while several accumulations of desmoplastic tissue are visible throughout the breast volume. The curved pink line in (**b**) indicates the skin margin, while the arrows in (**c**) indicate the skin involvement. The dashed square in (a) represents the crop region reported in Fig. 6.3

coronal. The sample has been scanned with an entrance air kerma of 8 mGy and its volume is approximately of 10 cm × 10 cm × 5 cm. From CT images the extension (maximum dimensions of 5 cm × 5 cm × 5 cm) and morphology of the tumor can be evaluated. Remarkably, the multiple-plane view enabled by tomography allows a clear evaluation of the various foci of the lesion, their connections and the skin involvement (see arrows in figure). These kind of features, which are cornerstones of therapeutic decision-making, are often difficult, sometimes impossible, to evaluate with standard imaging techniques.

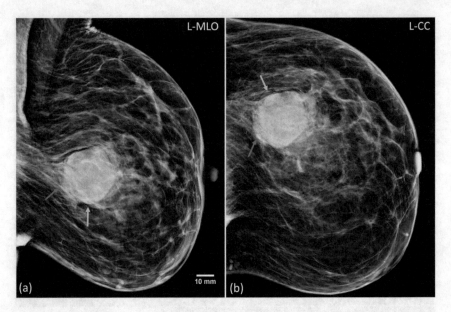

Fig. 6.2 Mammographic images of the patient before surgery corresponding to sample A: medio-lateral (i.e., sagittal) view (**a**) and cranio-caudal (i.e., transverse) view (**b**). Sharp margins of the opacity are indicated by yellow arrows while shaded margins are indicated by red arrows

For comparison, Fig. 6.2 shows the mammography performed few weeks before surgery. A high-density large round opacity (diameter of 4 cm) with some lobulations, surrounded by a non-homogeneous and non-specific less dense area, can be seen. While some of its margins are sharp (yellow arrows), others are shaded and difficult to interpret (red arrows) because of tissue superposition. By comparing the images, it is clear that by avoiding tissue superposition PBBCT allows a generally more accurate morphological description of the lesion, thus leading to a higher diagnostic confidence.

Of note, from the physical perspective, is the effect of phase retrieval on the visibility of fibroglandular details: panels (a), (b) of Fig. 6.3 show a zoom of Fig. 6.1 containing a thin fibroglandular spicula reconstructed without and with phase retrieval, respectively. Considering the line profiles in panels (c), (d), the fibrous detail is clearly visible only when the phase retrieval is applied, while, in the other case, it is well below the noise level.

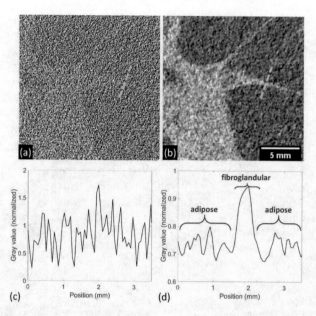

Fig. 6.3 Detail reconstructed without (**a**) and with (**b**) phase retrieval. In (**c**) and (**d**) profiles along the dashed lines of figures (**a**) and (**b**), respectively, are reported

6.2.2   Sample B

In Fig. 6.4 three orthogonal views of the sample B, acquired with an entrance air kerma of 7 mGy, are shown. The volume dimensions are 9 cm × 8 cm × 4 cm, while the crossing of line markers identifies the 2.5 cm × 2.5 cm × 2 cm tumor bulk. Surgical cuts performed during formalin fixation result in sharp interfaces between fibrous and adipose tissue and air gaps, which can be observed in the reconstruction. The tumor bulk embeds a hyper-dense sclerotic component and several microcalcifications (red circles). The irregularity of the lesion margin, as well as its spiculated appearance, are clearly visible, thus making the clinical picture compatible with a neoplastic lesion, which is confirmed by histological evaluation. Moreover, focusing on the large calcification (1.4 mm diameter) visible in the peripheral area of the sample (lower part of Fig. 6.4b), it is interesting to observe the presence of a cavity in its center, typical of benign rim calcifications.

To directly compare PBBCT and mammography, a slice oriented as the mammographic medio-lateral plane is chosen, using as a reference the aforementioned calcification, as marked by the circles in panels (a)–(c) of Fig. 6.5. It is clear that, while the mammographic image (c) represents an average of the attenuation properties of the 4 cm-thick compressed breast, the 60 μm thick CT slice (a) allows avoidance of tissue superposition. Furthermore, thanks to the three-dimensional nature of tomographic data, several processing operations other than averaging can be performed and, if needed, condensed in bi-dimensional images which are more common in breast imaging. As an example, in (b), the maximum intensity projection span-

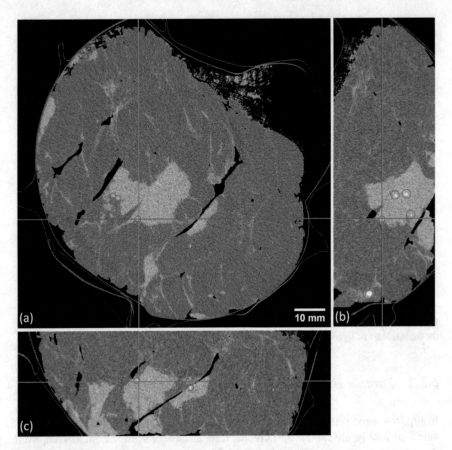

Fig. 6.4 Coronal (**a**), sagittal (**b**), transverse (**c**) views of the sample B. Line markers are centered in the bulk of the lesion, while red circles indicate microcalcifications

ning a thickness of 1.5 cm (about 300 slices) is reported. A generally good match with the mammography in terms of lesion's dimension and position is observed but, remarkably, dozens of microcalcifications in the tumor region are detected in the maximum intensity image, whereas they are missing in the mammographic examination. Moreover, following the maximum intensity projection operation, the large sclerotic component within the tumor is clearly visible (arrow in (**b**)).

In addition to orthogonal views display and bi-dimensional data reduction, CT images are suitable for 3D rendering as shown in Fig. 6.6. By adequately choosing the display thresholds, the fat tissue has been eliminated, fibroglandular/tumor structures have been made increasingly dark as a function of their density and the microcalcifications have been segmented (in red). The darker region within the tumor bulk encloses several calcifications and it identifies its hyper-dense sclerotic component. In general, 3D rendering has the advantage of capturing the global appearance of the lesion in terms of shape, distribution, extension and spiculation thanks to depth perception. Moreover, this kind of visualization enables further quantitative

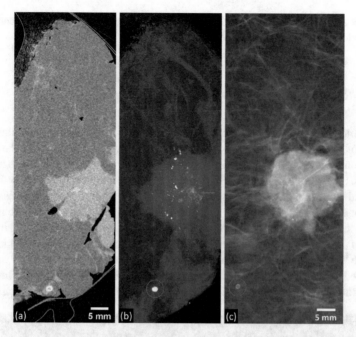

Fig. 6.5 Single slice (**a**) and maximum intensity projection (**b**) of the sagittal view of sample B. A crop of the medio-lateral pre-surgery mammographic image is reported in (**c**). Circles identify the benign rim calcification used as a reference while arrow in (**b**) indicates the hyper-dense sclerotic component

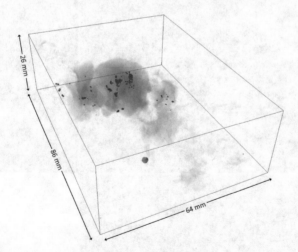

Fig. 6.6 3D rendering of the sample B. Increasingly darker regions represent fibroglandular/tumoral tissue with increasing density, red scattered volumes identify calcifications. The rendered volume is a sub-region of the whole scanned volume focusing on the lesion

analysis as, for instance, characterization of spatial and dimension distributions of microcalcifications and tumor modelling.

6.2.3 Sample C

The Sample C, of dimension $10\,cm \times 10\,cm \times 3\,cm$, is scanned with an entrance air kerma of 7 mGy. A multifocal lesion, marked by arrows in the image, can be seen in the coronal view displayed in panel (a) of Fig. 6.7, where the line markers are centered on a portion of it. In Fig. 6.8 a zoomed detail with dimension of $2.5 \times 2.5\,cm^2$ obtained from the tomographic scan (a) is compared with the respective histological image (b). From the PBBCT image a lesion with well defined smooth margins (yellow line) can be clearly distinguished from a contiguous structure with irregular margins (red line). This distinction is confirmed by matching the tomographic image with the histological examination, showing an encapsulated tumor (yellow line) and separated ductal structures with papillary lesion (red line). The light green line identifies a thickened skin tissue portion which has similar shape and orientation in both

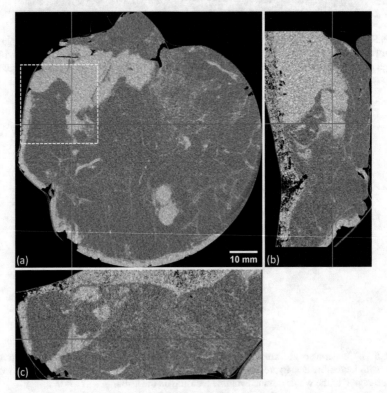

Fig. 6.7 Coronal (**a**), sagittal (**b**), transverse (**c**) views of the sample C. Line markers are centered on one portion of the largest lesion, while arrows indicate two different tumor foci. Dashed line encloses the detail shown in Fig. 6.8

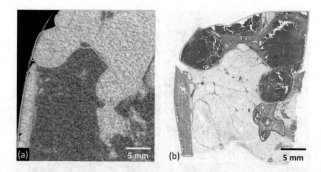

Fig. 6.8 Comparison between PBBCT (**a**) and histology (**b**). In both images, the region enclosed by the yellow line is an encapsulated lesion, the one within red line identifies ductal structures with a papillary lesion, the one within light green line is skin

PBBCT and histological images. It should be stressed that the matching between histological images and radiological images (with low radiation dose) is peculiar of the proposed PBBCT system. In fact, tissue superposition is encountered both in mammography and in tomosynthesis imaging, whereas insufficient spatial resolution generally affects other 3D techniques (e.g., MRI and ultrasound).

6.2.4   Future Developments

As demonstrated in Sect. 5.1, the propagation distance plays a crucial role in terms of image quality in PB configuration. Specifically, the SNR increase associated with the phase-retrieval is found to have, as a first approximation, a linear dependence with the propagation distance. As previously detailed, in light of these findings an extension will be installed at the SYRMEP beamline, enabling to reach patient-to-detector distances up to about 4.5 m. This upgrade is expected to improve the SNR by approximately a factor of two, even when both the small changes in magnification due to larger distance (from 1.05 at 1.6 m to 1.17 at 4.5 m) and the flux reduction (about 10% at 32 keV) due to the increased air attenuation are taken into account. To give an idea of the foreseen impact of the upgrade on image quality, sample C has been scanned with an exposure yielding a 2-fold higher SNR (i.e., 4-fold higher dose, 20 mGy MGD_t) and the results have been compared with the present 5 mGy image reference, as shown in Fig. 6.9. By zooming on a detail enclosing the margins of the main lesion (panels (c), (d)), it is clear that increasing the SNR by a factor of 2 allows to determine the presence of spiculae and thin connections of fibroglandular/tumoral tissue (green arrows), which are missed in the reference image (red arrows). Although the actual impact of the beamline upgrade should be assessed thorough dedicated measurements following its implementation, these results clarify the clinical impact of the upgrade, adding to the theoretical and quantitative demonstrations provided in Chap. 5.

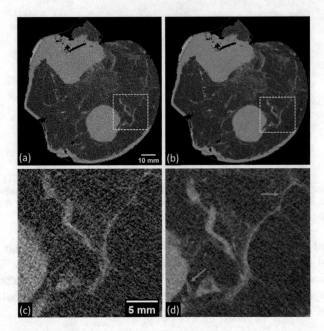

Fig. 6.9 Slice of the sample C scanned at MGD_t of 5 mGy and SNR = 9.1 (**a**), 20 mGy and SNR = 18.8 (**b**), mimicking a 5 mGy acquisition at the upgraded SYRMEP beamline. In (**c–d**) a zoom of a detail of (**a**), (**b**), as shown by the dashed line, is reported. Arrows point toward thin connections of fibroglandular tissue which are not visible in (**c**, red) and visible in (**d**, green). SNR is measured within the spheroidal hyper-dense mass

Along with the optimization of physical parameters such as the propagation distance, the use and/or development of ad-hoc reconstruction algorithms is a powerful tool to improve image quality, especially SNR, at a constant radiation dose. Albeit the presented images are reconstructed through the filtered-back-projection, which is arguably the most standard and widely used algorithm, the use of iterative reconstruction algorithms has demonstrated to provide convincing results [11]. In particular, the SYRMA-3D collaboration is developing a dedicated SART algorithm [12] making use of a 3D bilateral filter as a regularization factor during the iterative process [13, 14]. This algorithm, featuring several tunable parameters, has the advantage to allow for specific optimization on BCT images. Despite this optimization is still in progress, the beneficial effects of this algorithm are qualitatively shown in Fig. 6.10, displaying a section of sample C reconstructed with FBP (a) and SART (b) at 5 mGy of dose level. The use of dedicated SART reconstruction yields a SNR improvement of 40% (measured within the spheroidal lesion), while no evident degradation of spatial resolution is observed, as visible in the detail in panels (c), (d).

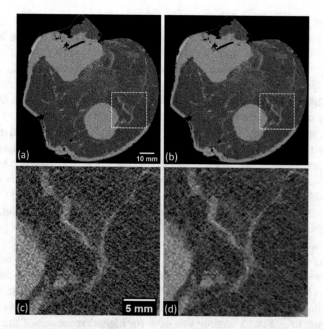

Fig. 6.10 Slice of the sample C reconstructed via FPB (**a**) and ad-hoc SART (**b**) algorithms. In (**c**), (**d**) a zoomed detail of (**a**), (**b**). as indicated by the dashed line, is shown. SNR measured within the spheroidal mass is 9.1 in (**a**) and 12.8 in (**b**)

6.3 Remarks Towards the Clinics

The images shown in this chapter represent an important step forward in the clinical implementation of phase-contrast breast CT at Elettra. The comparison between tomographic scans and standard mammographic images demonstrates that the 3D nature of tomographic data allows to avoid tissue superposition, while the high spatial and contrast resolutions determine a more accurate morphological description of neoplastic lesions. Clinically relevant conclusions on the malignant/benign nature, invasiveness and grading of a neoplastic lesion can be drawn from the detailed characterization of its volume, shape, margins, number and morphology of calcifications offered by PBBCT.

At present, digital breast tomosynthesis (DBT) is also envisaged as a tool to tackle issues as detection of microcalcifications and reduction of tissue superposition effects [15]. In fact, DBT is an emerging technology providing pseudo-3D reconstructions of the breast through the acquisition of multiple projections over a limited angular span. To better investigate its effectiveness in breast cancer screening and diagnosis, several DBT-based clinical trials are underway, but the reports regarding its capabilities in microcalcification detection are still mixed [16, 17]. In any case, diffrently from BCT which is a fully three dimensional technique, DBT offers pseudo-3D information, e.g., not allowing operations such as re-slicing in other view planes. Conversely, the possibility offered by CT of concentrating three-dimensional features in bi-dimensional images may be appealing to radiologists used to planar or quasi-planar mammographic techniques. As in the example shown in the previous section, the presence of hyper-dense structures and calcifications can be highlighted in a single image through maximum intensity operations, providing higher sensitivity if compared with conventional radiology. Moreover, being monochromatic CT inherently quantitative, the collected images can be used to characterize breast tissues in terms of absolute attenuation coefficients as shown, for instance, in [18]. The availability of high-resolution tomographic datasets also paves the way for 3D rendering and segmentation. This would be beneficial in evaluating spatial distribution of lesions and microcalcifications, and would serve as a reference for the surgeon in the pre-operative planning stage. Considering that PBBCT images can be matched to histological images even at low (i.e. clinically acceptable) radiation dose, their use in a clinical setting would allow for a more accurate tumor grading (TNM classification), where a precise assessment of lesion's dimensions is crucial. As in the case of specimens shown in this chapter, CT images can also serve as a guide in the specimen cutting process in pathological examination.

Focusing on clinical implementation, the exam duration is still a concern that needs addressing: to ensure patient comfort and to reduce motion-related artifacts, it should be kept as short as possible. To this end, the introduction of the flattening filter described in Sect. 5.3 would increase the usable vertical dimension of the beam (from 3.5 mm to $\gtrsim 5$ mm), leading to a 40% (or higher) reduction of the scan time. Concurrently, the possibility of reducing the number of projections along with the use of iterative reconstruction algorithms, as recently reported by [11], would further reduce the scan time by 20–30%, yet maintaining a comparable image quality. The combined effect of these improvements will bring to reduction of more than 50% in the overall exam duration.

References

1. Pacilè S, Brun F, Dullin C, Nesterets Y, Dreossi D, Mohammadi S, Tonutti M, Stacul F, Lockie D, Zanconati F et al (2015) Clinical application of low-dose phase contrast breast CT: methods for the optimization of the reconstruction workflow. Biomed Opt Express 6(8):3099–3112. https://doi.org/10.1364/BOE.6.003099
2. Longo R, Arfelli F, Bellazzini R, Bottigli U, Brez A, Brun F, Brunetti A, Delogu P, Di Lillo F, Dreossi D et al (2016) Towards breast tomography with synchrotron radiation at elettra: first images. Phys Med Biol 61(4):1634. https://doi.org/10.1088/0031-9155/61/4/1634
3. Baran P, Pacile S, Nesterets Y, Mayo S, Dullin C, Dreossi D, Arfelli F, Thompson D, Lockie D, McCormack M et al (2017) Optimization of propagation-based X-ray phase-contrast tomography for breast cancer imaging. Phys Med Biol 62(6):2315. https://doi.org/10.1088/1361-6560/aa5d3d
4. Keyrilainen J, Fernández M, Karjalainen-Lindsberg M-L, Virkkunen P, Leidenius M, Von Smitten K, Sipila P, Fiedler S, Suhonen H, Suortti P et al (2008) Toward high-contrast breast CT at low radiation dose. Radiology 249(1):321–327. https://doi.org/10.1148/radiol.2491072129
5. Zhao Y, Brun E, Coan P, Huang Z, Sztrókay A, Diemoz PC, Liebhardt S, Mittone A, Gasilov S, Miao J et al (2012) High-resolution, low-dose phase contrast X-ray tomography for 3d diagnosis of human breast cancers. Proc Natl Acad Sci 109(45):18290–18294. https://doi.org/10.1073/pnas.1204460109

6. Brun E, Grandl S, Sztrókay-Gaul A, Barbone G, Mittone A, Gasilov S, Bravin A, Coan P (2014) Breast tumor segmentation in high resolution X-ray phase contrast analyzer based computed tomography. Med Phys 41(11):111902. https://doi.org/10.1118/1.4896124
7. Brombal L, Golosio B, Arfelli F, Bonazza D, Contillo A, Delogu P, Donato S, Mettivier G, Oliva P, Rigon L et al (2018c) Monochromatic breast computed tomography with synchrotron radiation: phase-contrast and phase-retrieved image comparison and full-volume reconstruction. J Med Imaging 6(3):031402. https://doi.org/10.1117/1.JMI.6.3.031402
8. Pacilè S, Baran P, Dullin C, Dimmock M, Lockie D, Missbach-Guntner J, Quiney H, McCormack M, Mayo S, Thompson D et al (2018) Advantages of breast cancer visualization and characterization using synchrotron radiation phase-contrast tomography. J Synchrotron Radiat 25(5). https://doi.org/10.1107/S1600577518010172
9. Longo R, Arfelli F, Bonazza D, Bottigli U, Brombal L, Contillo A, Cova M, Delogu P, Di Lillo F, Di Trapani V et al (2019) Advancements towards the implementation of clinical phase-contrast breast computed tomography at elettra. J Synchrotron Radiat 26(4). https://doi.org/10.1107/S1600577519005502
10. Chen R, Longo R, Rigon L, Zanconati F, De Pellegrin A, Arfelli F, Dreossi D, Menk R, Vallazza E, Xiao T et al (2010) Measurement of the linear attenuation coefficients of breast tissues by synchrotron radiation computed tomography. Phys Med Biol 55(17):4993. https://doi.org/10.1088/0031-9155/55/17/008
11. Donato S, Brombal L, Tromba G, Longo R et al (2018) Phase-contrast breast-CT: optimization of experimental parameters and reconstruction algorithms. In: World congress on medical physics and biomedical engineering 2018. Springer, pp 109–115. https://doi.org/10.1007/978-981-10-9035-6_20
12. Kak AC, Slaney M, Wang G (2002) Principles of computerized tomographic imaging. Med Phys 29(1):107–107. https://doi.org/10.1118/1.1455742
13. Golosio B, Brunetti A, Cesareo R (2004) Algorithmic techniques for quantitative compton tomography. Nucl Instrum Methods Phys Res Sect B Beam Interact Mater Atoms 213:108–111. https://doi.org/10.1016/S0168-583X(03)01542-8
14. Oliva P, Golosio B, Arfelli F, Delogu P, Di Lillo F, Dreossi D, Fanti V, Fardin L, Fedon C, Mettivier G et al (2017) Quantitative evaluation of breast CT reconstruction by means of figures of merit based on similarity metrics. In: 2017 IEEE nuclear science symposium and medical imaging conference (NSS/MIC). IEEE, pp 1–5. https://doi.org/10.1109/NSSMIC.2017.8532786
15. Sechopoulos I (2013) A review of breast tomosynthesis. Part I, the image acquisition process. Med Phys 40(1). https://doi.org/10.1118/1.4770279
16. Marinovich ML, Hunter KE, Macaskill P, Houssami N (2018) Breast cancer screening using tomosynthesis or mammography: a meta-analysis of cancer detection and recall. JNCI J Natl Cancer Instit 110(9):942–949. https://doi.org/10.1093/jnci/djy121
17. Choi JS, Han B-K, Ko EY, Kim GR, Ko ES, Park KW (2019) Comparison of synthetic and digital mammography with digital breast tomosynthesis or alone for the detection and classification of microcalcifications. Eur Radiol 29(1):319–329. ISSN 1432-1084. https://doi.org/10.1007/s00330-018-5585-x
18. Piai A, Contillo A, Arfelli F, Bonazza D, Brombal L, Cova MA, Delogu P, Trapani VD, Donato S, Golosio B, Mettivier G, Oliva P, Rigon L, Taibi A, Tonutti M, Tromba G, Zanconati F, Longo R (2019) Quantitative characterization of breast tissues with dedicated CT imaging. Phys Med Biol 64(15):155011. https://doi.org/10.1088/1361-6560/ab2c29

Chapter 7
Do We Need Clinical Applications in Synchrotrons?

In general, the use of synchrotron radiation provides ideal working conditions for X-ray imaging which derive from the high flux, spatial and temporal coherence of the beam. On the other side, synchrotron light sources are huge facilities, limited in number, with high operational costs and infrastructural requirements. In other words, it can be questioned whether it is worth using a synchrotron facility for a given clinical imaging application or not. The answer to this question lies in the comparison between results obtained with SR and with more 'conventional' systems available in clinical or laboratory environments. In this context, the aim of the present chapter is to investigate the performances of two conventional systems with rather different application fields. In the next section, a first of its kind phantom-based comparison study between a clinically available BCT system and the SR PB imaging setup will be presented. In the second section the performances and possible applications of a state-of-the-art rotating-anode micro-CT system, capable of providing spatial and temporal coherence, are investigated. The last section will try to answer to the rather complex question kicking-off this chapter and it will provide a general overview of many existing or soon-to-come clinical applications of synchrotrons.

7.1 Synchrotron and Clinical BCT: A Comparison Study

In this section a direct quantitative and qualitative comparison between tomographic images of a breast-like phantom acquired by using both the SR setup and a clinical BCT machine in use at the Radboud University Medical Center (Nijmegen, The Netherlands) is presented, based on the results published in [1].

As discussed in Sect. 3.1, progresses in the development of BCT have been made in recent years and an increasing number of dedicated BCT systems with different acquisition modes (e.g., cone-beam, parallel-beam, helical) and detector types (e.g., flat-panels, photon-counting) have been proposed. In this lively context, there

© The Editor(s) (if applicable) and The Author(s), under exclusive license
to Springer Nature Switzerland AG 2020
L. Brombal, *X-Ray Phase-Contrast Tomography*, Springer Theses,
https://doi.org/10.1007/978-3-030-60433-2_7

is still a lack of image quality comparisons and no quantitative study performed among different systems, either based on conventional or synchrotron sources, has been published to date. Of course a higher image quality from synchrotron data is expected, but assessing the difference with clinically available systems can provide a benchmark on the current level of behaviour of SR-based techniques, and therefore establish its potential for clinical implementation. In other words, only showing that the gap with conventional techniques is substantial can provide justification for a SR clinical application.

This study makes use of both quantitative (objective) and qualitative (subjective) criteria. Specifically, signal-to-noise ratio (SNR), contrast-to-noise ratio (CNR), spatial resolution and noise power spectrum (NPS) are hereby used as indicators of image quality, possibly determining its diagnostic effectiveness. Namely, as discussed in previous chapters, SNR and CNR are related to low-contrast detail visibility (e.g., glandular tissue embedded in an adipose background), the shape of NPS reveals the image texture (i.e. low-frequency-peaked NPS are related to coarse image graininess; high-frequency-peaked NPS results in a finer grain noise) and spatial resolution determines the ability to detect small (high-contrast) details such as microcalcifications [2]. The comparison makes use of a breast-like phantom containing inserts mimicking relevant diagnostic features. The exposure parameters were automatically determined by the clinical BCT, while the SR irradiation parameters were tuned to replicate, as close as possible, the clinical conditions in terms of X-ray energy and delivered radiation dose.

7.1.1   BCT Dedicated Phantom and Experimental Setup

The used dedicated BCT phantom is shown in Fig. 7.1. It is produced by CIRS (model #12-685) and it has a semi-ellipsoidal truncated shape consisting of several slabs made of 100% breast-adipose equivalent material. A variety of targets are embedded into slab 9 as showed in panel (c): spheroidal masses of different diameters (1.80, 3.18, 4.76 and 6.32 mm) made of epoxy resin equivalent to breast carcinoma; cylindrical fibers of different diameters (0.15, 0.23, 0.41 and 0.60 mm); calcification clusters (CaCO3) of different grain sizes (0.13, 0.20, 0.29, 0.40 mm). The phantom was positioned at the system's isocenter both for the clinical and SR BCT setups.

The considered BCT clinical system is produced by Koning (Koning Corp., West Henrietta, NY) and it is installed at Radboud University Medical Center (Nijmegen, the Netherlands) [3]. A detailed description of the system can be found in literature [4–7], while only the most relevant features to this study are hereby reported. The system has a source-to-detector distance of 92.3 cm and a source-to-isocenter distance of 65.0 cm. The X-ray source is a rotating anode featuring a nominal focal spot size of 0.3 mm, whereas tomographic projections are acquired in half-cone beam geometry. The anode is made of tungsten while aluminum filtration is used to shape the energy spectrum. The tube is operated at a fixed voltage of 49 kV(peak), corresponding to a first half value layer of 1.39 mm Al (i.e. effective X-ray energy of

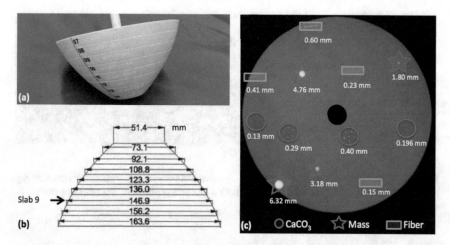

Fig. 7.1 Photograph of the phantom (**a**) and phantom dimensions in mm (**b**) (in mm). Details embedded in slab 9 (**c**): calcifications (CaCO3) in red circles, masses in blue stars and fibers in green rectangles

30.3 keV, evaluated from air kerma measurements after attenuation by various thickness of Al and using the weighted-energy average of a photon spectrum model as described in [8]). The X-ray source operates in pulse mode, with a constant 8 ms pulse length. A complete BCT acquisition consists of 300 projections over a full 360° revolution of the X-ray tube and detector in 10 s. The appropriate tube current is selected by acquiring two low-dose projections (16 mA, 2 pulses of 8 ms each per projection) images at right angles. The detector is a 39.7 cm × 29.8 8 cm flat-panel (4030CB, Varian Medical System, Palo Alto, California, USA) with a nominal pixel size of 194 μm. Tomographic reconstructions are performed according to the standard clinical workflow by using a Feldkamp-Davis-Kress (FDK)-based algorithm with a modified Shepp-Logan reconstruction filter, and an isotropic cubic voxel of 273 × 273 × 273 μm^3. The main components of the system are shown in panel (a) of Fig. 7.2a while panel (b) shows the phantom positioning. The automatically selected exposure parameters determine an air kerma of 13.5 mGy, corresponding to a mean glandular dose (MGD) value of 6.5 mGy.

The synchrotron-based images were acquired following the workflow described in Chap. 3: in order to match the clinical system conditions the energy was selected to be 30 keV, while 1200 projections were acquired in a 180° rotation delivering an air kerma of 14.2 mGy, corresponding to 6.7 mGy MGD. Prior to tomographic reconstruction, projection images are phase retrieved both with single- and two-materials approaches: as discussed in Sect. 2.5 the difference lies in the input δ/β values. Specifically, in case of single-material PhR $\delta/\beta = 2267$, corresponding to breast equivalent tissue is selected, whereas for the two-materials PhR $(\delta_1 - \delta_2)/(\beta_1 - \beta_2) = 795$ corresponding to a glandular/adipose interface is chosen.

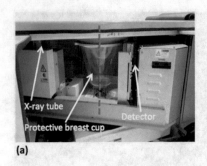 

(a)   **(b)**

Fig. 7.2 Photograph of the main breast CT system components (**a**). The red dotted line represents the system's rotation axis (i.e. isocenter). Isocenter position of the phantom during the measurements (**b**)

Since larger δ/β values correspond to smoother PhR filter kernels, the single- and two-materials approaches are hereinafter defined as smooth and sharp PhR kernels, respectively.

7.1.2 Image Quality Analysis

The CNR has already been defined in Eq. (5.20). Of note, the use of the standard deviation of the background to represent the magnitude of image noise, implies that the noise is assumed to be ergodic. With reference to the previous definition, CNR does not capture the dependence of detail visibility on the detail's size (i.e. Rose criterion). For this reason, the 'Rose' signal-to-noise-ratio (SNR_{Rose}) metric is introduced as [9, 10]:

$$SNR_{Rose} = CNR \times \sqrt{N_{pixel}} \tag{7.1}$$

where N_{pixel} is the number of pixel of the selected region of interest (ROI) within a given detail. Of note, this definition of signal-to-noise ratio has not to be confused with the SNR definition given in Chap. 5. Both CNR and SNR_{Rose} were evaluated for all the spheroidal masses shown in panel (c) of Fig. 7.1. As shown in panel (a) of Fig. 7.3, for each mass a circular ROI with a diameter scaling with the mass dimension was selected within the detail, while, for the background estimation, 10 evenly spaced ROIs were selected in the neighboring region. In the case of synchrotron-based datasets this analysis was repeated also by averaging 5 consecutive slices in order to match (as close as possible) the slice thickness of the clinical system, resulting in an effective voxel size of $57 \times 57 \times 250$ µm^3. With this choice a similar volume of a given detail is considered in each transverse slice for both systems.

While both CNR and SNR_{Rose} depend on the magnitude of the background noise, the image texture (or graininess) is characterized by the noise power spectrum (NPS), which is the noise spectral decomposition in the Fourier space. The in-slice NPS is

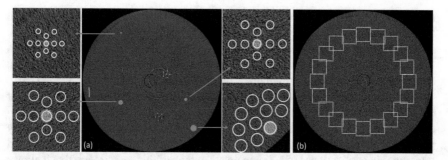

Fig. 7.3 SR-based tomographic reconstruction showing ROIs position for the CNR and SNR_{Rose} (**a**); ROIs position for the NPS evaluation in a homogeneous background are shown in (**b**)

a bi-dimensional map in Fourier space measured from a homogeneous phantom CT image by selecting equally sized ROIs and using the following definition [11, 12]:

$$\text{NPS}(u, v) = \frac{d_x d_y}{N_x N_y} \frac{1}{N_{\text{ROI}}} \sum_{i=1}^{N_{\text{ROI}}} |\mathscr{F}[I_i(x, y) - P_i(x, y)]|^2 \qquad (7.2)$$

where u, v are the spatial frequencies, d_x, d_y refers to the voxel size (in mm) along x and y dimension, N_x, N_y are the corresponding ROI dimensions measured in number of pixels, N_{ROI} is the number of selected ROIs, $\mathscr{F}$ denotes the bi-dimensional Fourier transform, $I_i(x, y)$ is the pixel value at position (x, y) of the ith ROI and $P_i(x, y)$ is a second order polynomial fit of $I_i(x, y)$. The subtraction with the polynomial term is a practical implementation of the de-trending procedure, aiming at removing any slowly-varying nonuniformities that may be caused from beam hardening effects, scattered radiation or nonuniform detector gain [12, 13]. As NPS is a spectral decomposition of image noise (σ), we have

$$\sigma^2 = \iint \text{NPS}(u, v) \, \mathrm{d}u \, \mathrm{d}v \qquad (7.3)$$

Following the procedure described by [12], in order to compare noise textures of images with different noise magnitude, the normalized NPS (nNPS) is defined as:

$$\text{nNPS}(u, v) = \frac{\text{NPS}(u, v)}{\sigma^2} \qquad (7.4)$$

In addition, since NPS maps of tomographic reconstructions usually show circular symmetry, it is common to show mono-dimensional radially averaged NPS curves making use of the identity $q = \sqrt{u^2 + v^2}$. The nNPS distributions, both bi- and mono-dimensional, were evaluated for both systems by selecting 20 evenly spaced square ROIs at a constant distance from the phantom center as shown in panel (b) of Fig. 7.3. Given the difference in the reconstructed voxel size between the two systems, the used ROIs have a 64×64 pixels area for the clinical and system 256 $\times$ 256 pixels area for the synchrotron datasets, meaning that each ROI represents a similar physical area for both systems. The uncertainty on radial nNPS curves was assessed by repeating the measure in 10 consecutive homogeneous slices and associating, for each spatial frequency, the corresponding standard deviation [13].

The spatial resolution of both systems was estimated directly from the images of the homogeneous portion of the phantom by using a novel approach recently introduced by Mitzutani and colleagues [14], which is based on a logarithmic intensity plot in the Fourier domain, and it has shown consistent results for both planar and tomographic applications [15]. The main advantage of this technique is that it allows to estimate spatial resolution directly from general sample images, not requiring dedicated phantoms, under the hypothesis of a Gaussian system point spread function (PSF). Although modern digital detectors, especially direct conversion devices, in general do not feature Gaussian response functions, the whole imaging chain PSF contains also the contribution of each processing step leading to the final tomographic image as detailed in Sect. 5.1.1.2. In particular, both the interpolation and apodization filter inherent to tomographic reconstruction contribute to smoothen the system PSF [16], usually described by a bell-shaped curve which, in case of the presented technique, is approximated by a Gaussian function. Under this assumption, the FWHM of the PSF can be determined from

$$\ln |\mathscr{F}_r [I(x, y)]| \simeq -\frac{\pi^2}{2 \ln 2} \text{FWHM}^2 |q|^2 + \text{constant} \tag{7.5}$$

where $\mathscr{F}_r$ is the radial Fourier transform. By performing a linear regression of the quantity $\ln |\mathscr{F}_r [I(x, y)]|$ as a function of $|q|^2$, yielding a correlation coefficient m, the FWHM can be easily estimated to be:

$$\text{FWHM} = \frac{\sqrt{-2 \ln 2 \times m}}{\pi} \tag{7.6}$$

Once the FWHM of the Gaussian PSF is known, the spatial resolution corresponding to the 10% of the modulation transfer function (MTF), measured in line-pairs per millimeter (lp/mm), can be easily estimated from [17]:

$$\text{MTF}_{10\%} (lp/\text{mm}) = \frac{1}{1.24 \times \text{FWHM(mm)}} \tag{7.7}$$

where the presence of the factor 1.24 is justified in the Appendix C. It should be remarked that, since not all the PSFs can be accurately approximated by a Gaussian function, this method cannot fully replace the direct PSF and MTF measurements based on line-patterns or small high-absorbing details, but has to be regarded as a fast and easy way to provide a spatial resolution estimate, possibly constituting a method for routine checks.

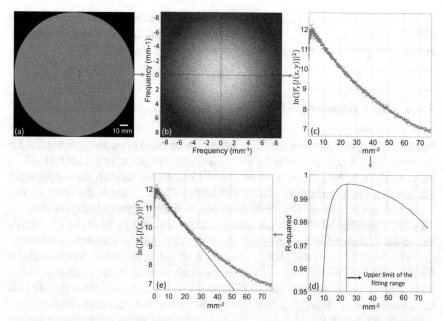

Fig. 7.4 Sketch of the implemented workflow for the estimation of spatial resolution. A detailed description of all the steps (**a**)–(**e**) can be found in text

As mentioned, this technique is rather new and not well established, so it is worth to report some practical details on its implementation. The scheme in Fig. 7.4 shows the implemented workflow for estimating the spatial resolution. A ROI comprised within an homogeneous portion of the phantom is selected (a) and the logarithm of the square modulus of its Fourier transform is computed (b). Then its radial average is plotted as a function of the spatial frequency squared (c). This plot should be fitted, in the region towards low spatial frequencies (squared), with a straight line [14, 15]. In order to identify the best fitting region, the fit procedure is repeated by finely varying the upper limit of the fitting interval and by plotting, as a function of the spatial frequency, its R-squared value (d). At this point, the fitting range yielding the maximum R-squared value is selected and the linear regression is plotted over the experimental data (e). In order to associate an uncertainty to the spatial resolution, the same procedure is repeated in 4 non-overlapping ROIs and the error is defined as the maximum difference among the spatial resolution estimates. Of note is that this procedure has been found to be robust, and compatible results are found by selecting different ROIs and/or different reconstructed slices. Moreover, it should be remarked that the PSF width is proportional to the square root of the regression coefficient, so that small inaccuracies in the fitting procedure translate in even smaller inaccuracies in the spatial resolution estimate (e.g.., a 10% error in the estimate of the regression coefficient corresponds to an error in the spatial resolution estimate of about 5%).

To complete the study of the two BCT setups, a qualitative analysis on the visibility of high-resolution details (i.e. calcification clusters and fibers) was performed by visually comparing the tomographic reconstructions of both systems.

7.1.3 BCT Image Quality Comparison: Experimental Results

Panel (a) of Figure 7.5 shows the CNR values as a function of the mass dimension for the two BCT systems (red color for the clinical and blue color for the SR system). In the case of SR images, the two phase-retrieval kernels and the two slice approaches (i.e. single slice and average over 5 consecutive slices to match the clinical slice thickness) are presented. The CNR in the clinical BCT system is higher than the SR case, regardless of the reconstruction and/or averaging methods: this is mainly due to the difference in the reconstructed voxel size. On the contrary, considering the detail visibility (i.e. the SNR_{Rose} metrics reported in panel (b)) which accounts for the number of pixels enclosed within the detail of interest, the synchrotron data show superior performances in all configurations, yielding, in case of the smooth PhR kernel and slice averaging, a 2.5–3 times higher SNR_{Rose} for all mass diameters.

Panels (a)–(c) of Fig. 7.6 show the bi-dimensional nNPS distributions for the clinical system and SR data with smooth and sharp PhR kernels. The noise in the clinical system is much coarser than in SR images as visible in the insets in the top-left corner of each panel. Given that, as expected, the bi-dimensional nNPSs have circular symmetry, their radial profiles were computed and plotted in panel (d). Peak frequencies largely differ when comparing the two systems, being 0.4 mm^{-1} for the clinical BCT, 0.9 mm^{-1} and 1.4 mm^{-1} for the synchrotron images reconstructed

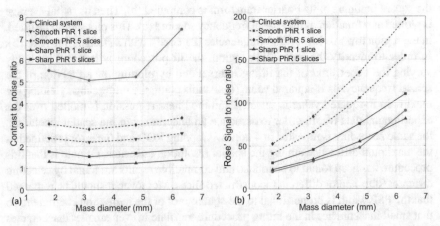

Fig. 7.5 CNR (**a**) and SNR$_{Rose}$ (**b**) as a function of mass dimension for clinical the breast CT (red solid line) and SR breast CT with smooth (blue dashed lines) and sharp (blue solid lines) phase-retrieval kernels

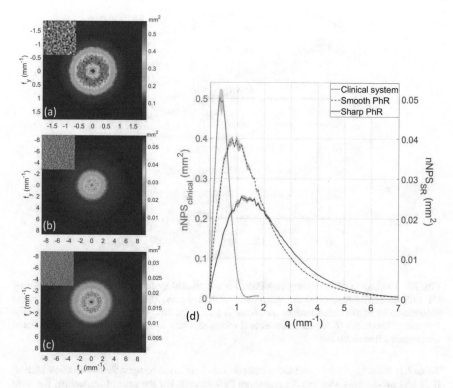

Fig. 7.6 Bi-dimensional nNPS for the clinical BCT system (**a**), synchrotron BCT with smooth (**b**) and sharp (**c**) PhR kernels. Of note, the extension of the frequency axis in (**a**) is different from (**b**) and (**c**). The inset in the top-left corner of each panel represents a 20×20 mm^2 homogeneous ROI. Radial averaged nNPS (**d**) for the clinical system (dashed red line) and SR BCT with smooth (dashed blue line) and sharp (solid blue line) phase-retrieval algorithm. Of note, the left y-axis refers to the nNPS of the clinical system while the right y-axis to the synchrotron data. The shaded region around each line represents one standard deviation uncertainty

with smooth and sharp PhR, respectively. In addition, the nNPS drops to 5% of its maximum value at 1 mm^{-1} for clinical BCT images, and at 5–6 mm^{-1} for SR datasets, meaning that the roll-off slopes of the nNPS curves are substantially different.

Following the procedure described in the previous section, the spatial resolution is estimated for all the different reconstructions as shown in Fig. 7.7: for each dataset a linear fitting region at small spatial frequencies is identified, where steeper linear fits indicate worse spatial resolutions. From the linear regressions the system resolutions were estimated to be 0.61 mm (FWHM) or 1.3 lp/mm (MTF$_{10\%}$) for the clinical BCT, 0.16 mm or 5.1 lp/mm for the smooth PhR and 0.12 mm or 6.8 lp/mm for the sharp PhR in SR images. The results of the quantitative analysis are summarized in Table 7.1.

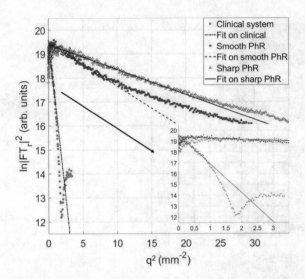

Fig. 7.7 Evaluation of the spatial resolution for the clinical system (red circles), and SR breast CT with smooth (blue squares) and sharp (blue-white triangles) PhR kernels. The logarithm of the absolute value of the radial Fourier transform is plotted as function of the square of the spatial frequency. The linear fit for each dataset is shown with black lines. The inset displays a zoom at lower spatial frequencies

Table 7.1 Summary of the quantitative analysis and comparison between the two systems: clinical BCT and SR datasets with smooth and sharp PhR kernels. For the sake of readability, the table reports the SNR_{Rose} and CNR values only for the 4.76 mm mass while, for the other masses, the quantitative values can be derived from Fig. 7.5. Where present, numbers enclosed within round brackets express the absolute uncertainty

	CNR	SNR_{Rose}	nNPS peak (1/mm)	FWHM (mm)	$MTF_{10\%}$ (lp/mm)
Clinical BCT	5.2	48	0.3	0.61 (0.02)	1.3 (<0.1)
Smooth PhR	2.3 (1 slice)	105	0.9	0.16 (<0.01)	5.1 (0.1)
	3.0 (5 slices)	135			
Sharp PhR	1.2 (1 slice)	55	1.4	0.12 (<0.01)	6.8 (0.1)
	1.7 (5 slices)	76			

Figure 7.8 displays the epoxy fibers for the clinical (a)–(d) and SR datasets with smooth (e)–(h) and sharp (i)–(l) PhR. All the fibers are visible in the SR breast CT regardless the PhR kernel, while the two smallest fibers (0.23 and 0.15 mm in diameter) are not distinguishable in clinical BCT images. Figure 7.9 shows image details of the calcification clusters for the clinical (a)–(d) and SR datasets with smooth (e)–(h) and sharp (i)–(l) PhR. For the clinical BCT system, no calcification cluster

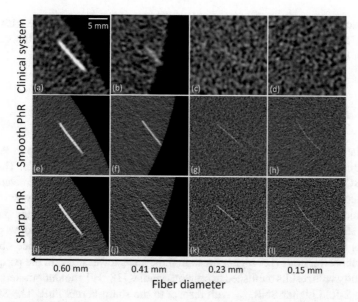

Fig. 7.8 Details of the epoxy fibers reconstructed (**a**)–(**d**) with the clinical BCT system, (**e**)–(**h**) smooth and (**i**)–(**l**) sharp PhR kernels for the SR BCT

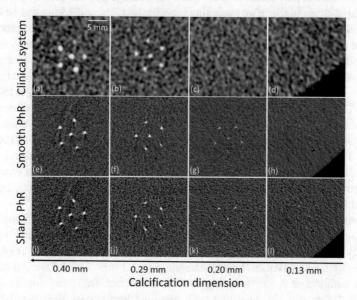

Fig. 7.9 Details of the calcification clusters reconstructed (**a**)–(**d**) with the clinical BCT system, (**e**)–(**h**) smooth and (**i**)–(**l**) sharp PhR kernels for the SR BCT

with diameter below 0.20 mm can be properly identified, while in the case of SR breast CT the smallest calcification (0.13 mm in diameter) represent the visibility limit for both the smooth and sharp PhR kernels.

7.1.4 BCT Image Quality Comparison: Discussion

From the data presented in the previous section it is clear that the gap in terms of image quality between clinical and SR breast CT systems is quite wide. The CNR in SR BCT images is found to be almost constant at different mass diameters, with small fluctuations mainly due to different noise levels. In particular, the two masses (dimensions of 3.18 and 4.76 mm) positioned closer to the center of the phantom show a slightly lower CNR with respect to the two located in the phantom's periphery: this behavior is compatible with the usual radial noise dependence observed in CT reconstructions (i.e. higher noise in the center, lower noise in the periphery). Coherently with results published in previous studies [18, 19], the smooth-kernel PhR yields a 2-fold higher SNR_{Rose} with respect to the sharp-kernel PhR. The SNR_{Rose} for the SR setup can be up to 3-times higher with respect to the clinical BCT if the smooth reconstruction kernel is used when the average of 5 slices is considered, or more than 2-times higher if no averaging is performed. This difference can be mainly attributed to the high-efficiency and low-noise of the photon-counting detector, to the presence of phase-contrast effects, and the subsequent application of phase-retrieval filter, and to the higher dose-efficiency of the synchrotron system due to the beam monochromaticity. In addition, thanks to the laminar shape of the beam and the large isocenter-to-detector distance, the SR setup allows to obtain inherently scatter-free images. Considering SR-based data, it should be noted that, if the noise of each slice was uncorrelated, the expected SNR_{Rose} and CNR increase due to the averaging of 5 slices would be of a factor $\sqrt{5}$, whereas the observed factor is much smaller (between 1.3 and 1.4). This is mainly related to the application of the phase-retrieval which, being a 2D filter in the projections domain, introduces a certain degree of correlation also between neighbouring pixels belonging to different rows of pixels, hence to different slices.

The nNPS evaluation revealed that the synchrotron images have a 3 to 5 times higher peak frequency (for the smooth and sharp PhR kernels, respectively) and a generally shallower roll-off slope, meaning that the contribution to the image noise is not negligible up to 6 mm^{-1}, to be compared with 1 mm^{-1} of the clinical system's case. In addition, it is worth noting that the NPS peak frequency for the clinical BCT, i.e. 0.4 mm^{-1}, is consistent with previous findings by Betancourt-Benitez and colleagues [7], who characterized the system before its commercialization. The observed differences in terms of nNPS between clinical and synchrotron data reveals that the SR setup imaging chain (i.e. detector, image processing and tomographic reconstruction) provides generally sharper or, equivalently, less correlated noise: this is ultimately related to the smaller detector pixel size and to the higher image-sharpness offered by direct-conversion photon-counting detectors.

Despite being a model containing several simplifications (e.g., the PSF is assumed to be constant and Gaussian throughout the image) not allowing a detailed description of the system PSF (e.g., resolutions in radial and tangential directions cannot be uncoupled), the spatial resolution assessment through images of the homogeneous phantom has been proven to a robust and easy-to-implement technique. In facts, the results obtained on the SR images, with both the smooth and sharp PhR kernels, are compatible with conventional spatial resolution estimates (based on the edge spread function technique) documented in Chap. 5 and in other studies [19–21]. Quantitatively, the spatial resolution of the SR system was found to be 4 to 5 times better than the clinical system (5–7 lp/mm for the synchrotron to be compared with 1.3 lp/mm for the clinical setup). Interestingly, synchrotron images outperform every clinical breast CT setup reported in literature so far in terms of spatial resolution, the maximum being 5 lp/mm for a photon-counting breast CT system proposed by Kalender and co-workers [22–24]. The qualitative analysis in terms of detail visibility showed that both the smallest fibers (i.e. diameter of 0.15 mm) and calcification clusters (i.e. diameter of 0.13 mm) can be detected in the SR-based images, while details with dimension in the order of 0.20 mm or below cannot be properly identified in the clinical BCT system. As mentioned in the previous chapter, the correct detection of such details plays a crucial role in the diagnostic process since both the presence of microcalcifications and spiculae (i.e. small fibers protruding from a bulk mass) are signs of malignancy.

Before concluding this section, it should be remarked that the implementation of SR BCT to the clinical realm presents also some practical drawbacks, the main being the longer scan time with respect to clinical systems due to the limited vertical dimension of the beam, to the need for patient rotation and to the limited detector readout speed. This can lead to motion artifacts due to both voluntary and involuntary movements of the patient, possibly impairing image quality (mainly spatial resolution). This issue has been encountered also in a clinical context suggesting the use of a breast immobilizer [25]. As mentioned in Chap. 6, the SYRMA-3D collaboration is devoting several efforts towards the reduction of the scan time, while the usefulness of immobilization systems is being investigated.

7.2 Monochromatic PB Micro-CT with a Rotating Anode Source

In the previous section a comparison between a synchrotron and a (conventional) clinical system was performed focusing on a specific imaging application, i.e. BCT. The two systems largely differ in terms of geometry, detector and, most importantly, X-ray quality, where the SR spatial and temporal coherence provide the key advantage over the clinical BCT. On the other hand, compact laboratory setups (as opposed to SR setups) based on conventional X-ray sources enabling monochromatic phase-contrast imaging exist, even if their application usually focuses on small samples (i.e. in the millimeter scale) due to the limited field of view and/or limited flux. These limitations impose *a fortiori* a shift from clinical to preclinical or nonclinical studies, often based on *ex-vivo* samples. Nonetheless, the higher contrast or contrast sensitivity offered by phase-sensitive techniques when imaging soft samples, represents a key advantage over attenuation imaging.

In this section, a monochromatic PB micro-CT system based on a state-of-the-art rotating anode source is presented, reporting a detailed characterization, both in planar and tomographic configurations, and applications to two biological samples of medical interest. In addition, some practical considerations on possible trade-offs between scan time and image quality as well as improvements on the presented setup are discussed. All the experimental work hereby presented has been carried out at the X-ray Phase Contrast Imaging laboratory of the Department of Medical Physics and Biomedical Engineering of University College London (London, UK) and partly described in [26].

As discussed in Sect. 2.2, over the last two decades, many phase-sensitive techniques have been developed (e.g.., propagation-based, analyzer-based, edge-illumination, interferometric etc.) and most of them are in use with synchrotron and, in some cases, conventional sources [27–30]. As mentioned, propagation-based imaging is, in terms of experimental setup, the simplest to implement as in principle it does not require optical elements or multiple exposures. On the other hand, in terms of X-ray source characteristics, PB has more stringent requirements, demanding for high spatial coherence and, especially at small magnifications, high detector spatial resolution. For this reason, most of its applications have been so far limited either to synchrotron radiation facilities or to low-power micro-focal sources [28, 31–34]. In this context, the development of compact and partially coherent high-flux X-ray sources is an active area of research [35, 36].

Several laboratory X-ray sources, based either on liquid-metal, fixed or rotating targets, are capable of producing sufficient flux and spatial coherence to be used for phase-contrast imaging purposes, the main advantages over synchrotrons being availability, compactness and low costs [37–40]. Moreover, monochromator crystals selecting the characteristic X-ray lines can be coupled to the source, thus producing quasi-monochromatic spectra. It is noteworthy that, albeit not being essential for PB imaging, the use of narrow monochromatic radiation is advantageous even when no dose-efficiency constraints are present, as it allows performing a straightforward quantitative analysis and avoiding beam hardening effects.

In the following, the theoretical background presented in Chap. 2 will be widely used to characterize the system in terms of spatial resolution, coherence, quantitative-ness, stability, and contrast sensitivity. Planar and tomographic images of custom-built wire phantoms are compared with theoretical predictions. In addition, the applications on two biological samples of medical interest demonstrate the feasibility of monochromatic PB imaging μ-CT with laboratory-compatible exposure times from tens of minutes to hours.

7.2.1   *System Characterization*

A schematic overview of the experimental setup is given in Fig. 7.10. X-rays are produced by a Rigaku Multi-Max 9 rotating anode source, featuring a copper anode and
operated at 46 kV(peak) and 26 mA corresponding to a power of 1.2 kW. The source
is coupled to a double bent multilayer VariMax Cu-HF monochromator, providing an
energy resolution of about 1% at 8 keV (copper k_α emission lines) and focusing the
beam to a 210 μm focal spot [41, 42]. The source dimension is defined by a golden
plated pinhole collimator with a diameter of 75 μm, located at the focus position of
the monochromator. This arrangement (i.e. monochromator and collimator) results
in an integrated flux of about 10^8 ph/s and a divergence of 5 mrad. The sample was
positioned at 88 cm from the source, while the propagation distance was set to 11 cm,
corresponding to a magnification of M = 1.13. At this distance, the field of view was
diamond shaped with dimensions of about 5×5 mm^2. The sample alignment and
rotation were performed through a piezometric motor stack with 5 degrees of freedom
and sub-micrometric precision. The imaging detector was a charge-coupled device
(CCD) camera featuring a 4.54 μm pixel size, coupled through a fiber-optic plate
to a Gadox scintillator (Photonic Science). Both the detector PSF and the source
intensity distribution were measured with the slanted edge technique by using a 50
μm thick lead blade, the unsharpness and finite-thickness effects of which can be
neglected given the system energy and spatial resolution [43]. The absorbing edge
was placed alternatively close (distance of 10 cm) to the source and in contact with the
detector to provide independent measurements of the source dimension and detector
PSF, respectively. As a cross-check, the blade was also positioned at sample position
yielding, by taking into account the magnification, consistent results.

The overall spatial resolution of the system is the key parameter in determining
whether or not phase effects can be observed. Therefore, the overall system PSF was
evaluated as:

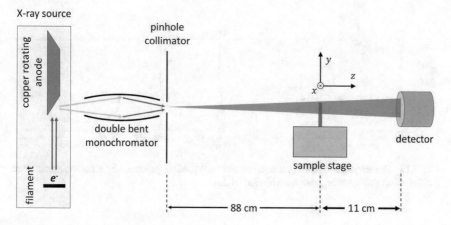

Fig. 7.10 Schematic overview of the experimental setup

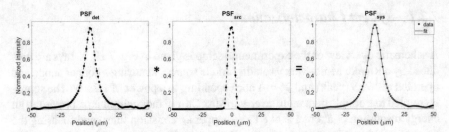

Fig. 7.11 Detector (left), source (center) and system (right) PSFs projected at the sample position. The system PSF has been fitted (red solid line) with a linear combination of Lorentzian and Gaussian functions

$$\text{PSF}_{\text{sys}}(x, y; M) = \text{PSF}_{\text{det}}(Mx, My) * \text{PSF}_{\text{src}}\left(\frac{M}{M-1}x, \frac{M}{M-1}y\right) \quad (7.8)$$

where this expression is analogous to Eq. (2.13) computed at sample position instead of detector position. In Fig. 7.11 the measured detector PSF (left), source distribution (center), and their convolution (right) are reported as a function of the spatial coordinate at the sample position according to Eq. 7.8. The experimental system PSF has been fitted with a linear combination of a Lorentzian and a Gaussian function. The blurring due to the detector response is of 12 μm full-width-half-maximum (FWHM), while the source size projected at the sample position is of about 10 μm, resulting in an overall resolution of about 14 μm FWHM.

Given the system PSF, the intensity profile given by a wire of known composition can be theoretically calculated according to Eq. (2.8), where the refraction angle produced by a cylinder (i.e. wire) oriented along the y direction can be analytically expressed as:

$$\alpha(x) \simeq \frac{2\delta x}{\sqrt{r^2 - x^2}} \quad (7.9)$$

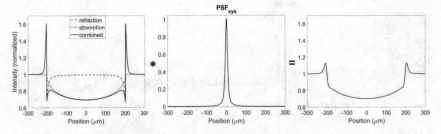

Fig. 7.12 Theoretical refraction, attenuation and total profiles produced by a homogeneous wire (left), system PSF (center), their convolution (right)

In the left panel of Fig. 7.12 the refraction (blue line), transmission (red line), and total (black line) intensity profiles calculated according to Eq. 2.12 are reported. Despite the smearing due to the convolution with the system PSF (central panel), the expected signal (right panel) still shows edge-enhancement contrast, indicating that the system spatial coherence and spatial resolution are sufficient to detect phase effects.

7.2.2 Acquisition Parameters and Data Processing

Two *ad-hoc* built wire phantoms have been imaged in planar and tomographic geometries, respectively. The planar acquisition was performed with an overall exposure time of 100 s whereas the long exposure CT-scan was acquired over 1440 projections with an exposure time of 10 s per projection, corresponding to a total exposure time of 4 hours. The tomographic scan has been repeated with a 20 times shorter exposure time (i.e. fast scan), acquiring 720 projections with an exposure of 1 s, resulting in a total exposure of 12 minutes. Similarly, in the long scans, the biological samples have been imaged with the same number of projections and an exposure time of 6 s per projection, corresponding to a total exposure of 2.4 hours, whereas the short scan has been obtained by reducing the exposure of a factor of 10, i.e. acquiring 720 projections of 1.2 s each, resulting in a total exposure of 14 minutes.

The planar data were processed by a conventional dark current subtraction and flat field normalization, whereas for CT scans the projections have been normalized using a dynamic flat field approach based on the principal component analysis of the flat images to compensate for beam intensity variations over long exposures [44]. The normalized projections have been (optionally) phase-retrieved and reconstructed through the same reconstruction software used to process synchrotron-based data [45], as detailed in Sect. 3.7. Of note, the reconstruction has been performed assuming a parallel beam geometry irradiation since, considering the small sample sizes and setup geometry, the beam divergence within the sample was smaller than the system spatial resolution, thus not requiring the use of a cone beam reconstruction.

7.2.3 Plastic Phantoms

Both the wire phantoms consisted of 3 different high-purity plastic rods made of Polybutylene terephthalate (PBT), Polyethylene terephthalate (PET), and Nylon. The real and imaginary parts of the refractive index used for the theoretical calculations are listed in a publicly available database [46] and are reported in Table 7.2.

The first test of the system quantitativeness was performed by imaging a planar phantom consisting of 3 vertically oriented wires made of PBT, PET and nylon, plus 1 horizontal PBT wire (panel (a) of Fig. 7.13). For each of the vertical wires, a line

Table 7.2 Physical properties of the wires used for the phantoms

	$\delta \times 10^{-6}$	$\beta \times 10^{-9}$	δ/β	Density (g/cm^3)	Diameter (μm)
PBT	4.45	9.79	454	1.31	180
PET	4.70	11.1	423	1.40	400
Nylon	3.99	7.25	550	1.13	160

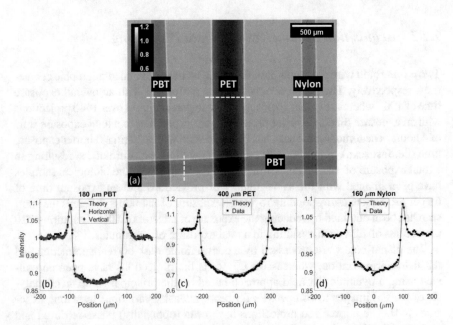

Fig. 7.13 Image of the planar wires phantom (**a**) and plots of the intensity profiles (**b**)–(**d**) along the white dashed lines. The image results from dark current subtraction and flat field normalization

intensity profile is compared against their respective theoretical profiles, accounting for the nominal values of density, attenuation and refraction of each material (b)–(d). The overall agreement between theory and experimental data is remarkable both considering phase and attenuation contrast, the largest discrepancy being a slight underestimate ($< 5\%$) of the PET attenuation. Moreover, by comparing profiles extracted from both the horizontal and vertical PBT wires (b), the same phase sensitivity is achieved in both directions due to the circular symmetry of the source.

Wires of same materials and sizes were used to assess the system performances in CT acquisitions. In panel (a) of Fig 7.14, a tomographic slice of the long scan is shown: thanks to the beam monochromaticity the reconstruction is inherently quantitative, thus, far from the sample boundaries where the edge-enhancement effect is present, the gray level represents the linear attenuation coefficient. To obtain the theoretical profiles for the CT case, a sinogram composed by a set of identical line profiles was created for each wire and then reconstructed following the same workflow used

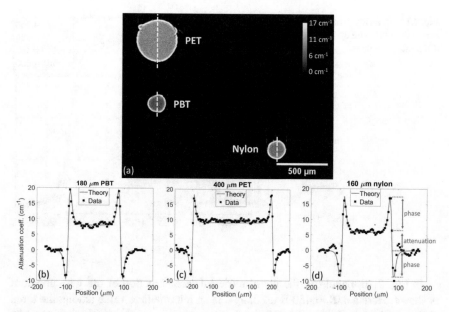

Fig. 7.14 Reconstructed slice of the wire phantoms (**a**) and plots of the intensity profiles (**b**)–(**d**) along the white dashed lines. The structure visible in the top corners of (**a**) is part of the cylinder that was used to keep the phantom in place

for the experimental data. As for the planar image, a good agreement is observed when comparing theoretical and experimental profiles across the wires for both phase and attenuation signals, except for a small discrepancy ($<10\%$) in the attenuation coefficient of PET (b)–(d). The fact that the refraction fringes (i.e. edge-enhancement signal) are well matched by the theoretical predictions for a scan acquired over several hours, provides an indirect assessment of the system stability and piezometric motors reproducibility: vibrations or spatial drifts of the source, sample or detector, or slight inaccuracies in the sample repositioning after the periodic flat field images acquisition, would result in a broader effective PSF, thus smearing out the fringes. Furthermore, by defining the refraction (or phase-contrast) signal as the sum of the overshoots of dark and bright fringes (see panel (d)), this is in all cases between 1.5 and 3 times higher than the attenuation signal.

As discussed in Sect. 2.4, CT projections were processed by applying the Paganin's single shot phase-retrieval algorithm. In order to adequately choose the filter parameter, it is common practice to tweak δ/β until refraction fringes disappear without introducing an excessive smoothing. Such a procedure is often applied when dealing with polychromatic X-ray spectra or with samples of unknown composition. To demonstrate this practice, several profiles taken across the PBT wire are shown in Fig. 7.15. Each profile has been reconstructed using a δ/β value in the range 250–550: thanks to the beam monochromaticity, it is found that the optimal δ/β is 450 that well matches its nominal value (see Table 7.2).

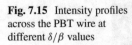

Fig. 7.15 Intensity profiles across the PBT wire at different δ/β values

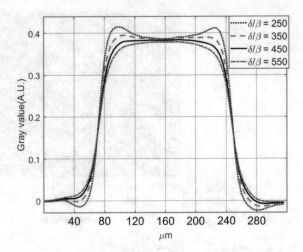

In Fig. 7.16, panel (a), the phase retrieved reconstruction of the wire phantom is shown. Here a $\delta/\beta = 450$ is used, as it is an intermediate value among the three different plastics. As expected, the refraction fringes are no longer visible, while the noise has been significantly suppressed due to the 'low-pass filtering' effect of phase-retrieval detailed in Sect. 5.1. This can be clearly appreciated in the gray level histograms in panels (c), (d), which are obtained by selecting circular ROIs at the center of each wire for both the images with and without the phase retrieval: after phase retrieval the three materials can be easily separated based on the gray values of each voxel. The ROIs are selected far from edges where the gray level distribution is flat and have equal areas to provide histograms with equal statistics. Given the major increase in contrast sensitivity achieved with the phase retrieval, it is interesting to observe the results obtained from the same sample scanned with a 20-fold shorter exposure time, as shown in panel (b). Even though a broadening of the distributions due to the reduced statistics can be seen, the histogram in panel (e) shows that the materials are still clearly distinguishable. In quantitative terms, we observe that the central values of the gray level distributions are separated, respectively, by ~25 standard deviations for the long and ~10 for the short exposure scans. This clear separation, between materials of similar attenuation properties, is advantageous in all those applications involving subsequent data processing steps such as segmentation.

The quantitative results extracted from tomographic images are summarized in Table 7.3. For all materials, the measured attenuation coefficient is compatible, within the noise fluctuations, with the theoretical values; the maximum discrepancy in terms of mean value is observed for PET wire and it is smaller than 10%. This result is compatible with the findings of the planar image where PET has been found to be more absorbing than its nominal value. To estimate the effects of phase retrieval, the contrast with respect to the least absorbing material, i.e. Nylon, has been measured both before and after the application of the retrieval algorithm. As expected, no significant differences in the detected contrast are observed, indicating that the image

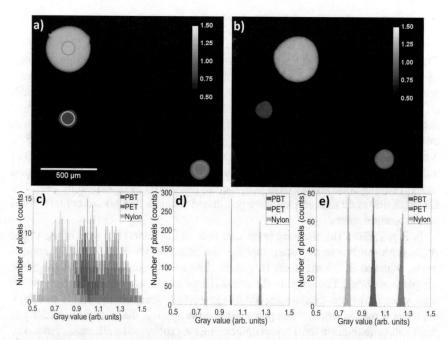

Fig. 7.16 Image of the wire phantom after phase retrieval for the 4 hours long exposure (**a**) and the 12 minutes long exposure (**b**). In (**a**) the ROIs used for the histograms are reported. Gray level histograms are relative to the wires phantom reconstructed without (**c**, see also Fig. 7.15) and with phase retrieval for the long (**d**) and short (**e**) exposures

Table 7.3 Quantitative results obtained from CT reconstructions. C–no phrt, C–phrt and C–phrt short refer to the contrast of long exposure non-phase-retrieved, phase-retrieved and short exposure phase-retrieved acquisitions, respectively, whereas subscript th and exp refers to theoretical and experimental values, respectively. Uncertainties are computed by following standard error propagation rules

	μ_{th} (cm^{-1})	μ_{exp} (cm^{-1})	rel error (%)	C–no phrt (%)	C–phrt (%)	C–phrt short (%)
PBT	7.98	7.8±0.8	−1.8	30±18	27.9±0.7	28.5±2.0
PET	9.01	9.8±0.8	8.7	62±20	59.9±0.8	59.9 ±2.2
Nylon	5.91	6.0±0.8	2.0	–	–	–

retains its quantitativeness (see Sect. 2.6). On the contrary, a major improvement in the contrast sensitivity (i.e. the associated uncertainty), going from about 20% to values smaller than 1%, is found. Also when the short exposure acquisition is considered, the contrast sensitivity is still around 2%, clearly sufficient for material differentiation, while no contrast variation is observed.

7.2.4 Biological Samples

The scans of two biological samples were acquired to assess the imaging potential of the experimental setup on complex objects. The first sample is an esophageal acellular matrix (ACM), derived from a piglet, provided by Institute of Child Health (ICH). The ACM was derived via an established decelluarization technique named detergent enzymatic treatment (DET) [47, 48]. Following the DET the sample was critical point dried using CO_2. The sample has an approximate size of $5 \times 5 \times 3$ mm^3. The second sample is a lobe (dimension approximately $3 \times 5 \times 3$ mm^3) of a dehydrated fibrotic murine lung generated from bleomycin-induced lung fibrosis model (sample collected 28 days post-bleomycin, 25IU) as described by [49]. For CT acquisitions all the samples were positioned within a thin plastic cylinder fixed on the rotation stage.

In Fig. 7.17(a), (b), the long (exposure time of 2.4 hours) CT scan of the piglet ACM is shown before and after applying the phase retrieval ($\delta/\beta = 100$), respectively, whereas in (c) the short (exposure time of 14 minutes) scan of the same sample is reported. Focusing on the detail shown in panels (d–f), it is clear that the high noise in the non-phase-retrieved image possibly hampers the ability to differentiate soft tissues while, when phase retrieval is applied, the contrast sensitivity is sufficient to distinguish the 4 layers composing the esophageal wall, namely mucosa,

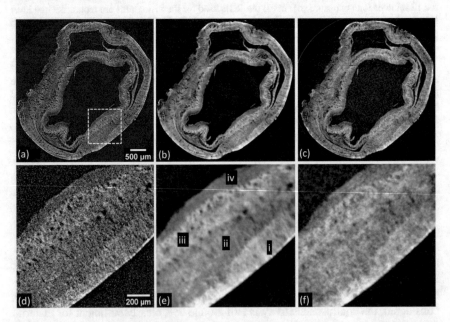

Fig. 7.17 Decellularized piglet esophagus scan with long exposure without (**a**), (**d**) and with (**b**), (**e**) phase retrieval, and short exposure with phase retrieval (**c**), (**f**). The dashed square in (**a**) represent the detail zoomed-in in the lower panels. The labels in (**e**) identify from right to left the adventitia (i), muscularis propria (ii), sub-mucosa (iii) and mucosa (iv)

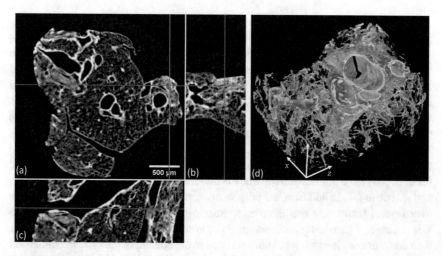

Fig. 7.18 Trans-axial (**a**), sagittal (**b**), transverse (**c**) slice and 3D rendering (**d**) of the fibrotic mouse lung sample

sub-mucosa, muscularis propria and adventitia. Remarkably, despite a higher noise level, the tissue layers are distinguishable also in the short exposure scan as visible in panel (e).

Panels (a)–(c) in Fig. 7.18 show the orthogonal views of the mouse lung sample phase-retrieved reconstruction ($\delta/\beta = 50$), while in panel (d) the 3D rendering is reported. Dense fibrotic tissue can be distinguished in the sub-pleural peripheral and bronchovascular regions, as shown for instance at the markers crossing position, with bronchi and bronchioles a prominent feature in the 3D rendering. Quantification of changes in parenchymal density, as seen in fibrosis, or measurement of airway or vascular remodelling represent potential pre-clinical applications of this imaging technique.

7.2.5 Remarks and Outlooks on High-Power Rotating Anode PB Systems

Most of laboratory phase-contrast imaging setups are based on polychromatic, low power, microfocal sources and cone beam scan geometries (i.e. large beam divergence) featuring high magnifications. Conversely, the results reported in this section show that quantitative PB imaging can be attained also by using compact high-power rotating anode sources which, coupled with dedicated optics, are capable of providing high-flux and temporal coherence. The geometry of this system resembles, in some way, the irradiation geometry commonly found in synchrotron facilities, where small magnifications and parallel beam reconstruction are used.

Specifically, the described setup can be appealing for light materials, such as plastics or soft tissues, with dimensions in the millimeter scale requiring high contrast sensitivity and spatial resolution in the order of 10 μm, while scan times range from hours to tens of minutes. The system, capable of delivering an integrated flux of 10^8 ph/s, has been characterized in terms of spatial coherence and detector spatial resolution, resulting in an overall PSF at the sample position of 14 μm FWHM: this value represents an optimal trade-off between spatial coherence and X-ray flux since the source size projected at the sample position is comparable to the detector PSF. The comparison between experimental data and theoretical prediction allowed to demonstrate the quantitativeness of the system, as an overall good agreement is found for both phase and attenuation signals, the maximum difference being <5% in planar and <10% in CT. In addition, the proposed setup has proven to be sufficiently stable over several hours, that was the time to acquire the high-statistics CT scans, while it is capable of providing a refraction (i.e. phase-contrast) signal 2–3 times higher than conventional X-ray attenuation. As done in the context of the synchrotron-based BCT project (see Chap. 5), the effects of the phase-retrieval algorithm on image noise and contrast sensitivity have been examined, showing that a 20-fold improvement in contrast sensitivity (from ∼20% to ≲1%) is achieved for the wire-phantom CT scan. This opens up the possibility of significantly reducing the exposure time: going from 4 hours to 12 minutes, contrast resolutions around 2% are found, still providing a fine resolving power between different soft materials. The tests on two biological samples of medical interest have shown the potential of the system in the field of pre-clinical applications as, for instance, digital histology or some aspects of regenerative medicine such as tissue/scaffold interactions, involving samples with dimensions in the millimeter scale.

As a general remark it is worth noting that, in addition to the configuration reported in this study, the setup is inherently flexible as it allows adjusting the spatial coherence, by replacing the pinhole collimator defining the source size, and the magnification. Moreover, by inserting a vacuum pipe to prevent air attenuation, the field of view can be in principle enlarged at a constant fluence rate. In fact, keeping the spatial coherence constant, the linear source size d (i.e. the collimator diameter) can be scaled with the source-to-detector distance $z_0 + z_1$, thus compensating the fluence rate reduction due to the larger source-to-detector distance by the larger dimension of the source:

$$\text{fluence rate}\left(\frac{\text{photons}}{\text{mm}^2 s^1}\right) \propto \frac{d^2}{(z_0 + z_1)^2} \propto \frac{(z_0 + z_1)^2}{(z_0 + z_1)^2} = \text{constant} \qquad (7.10)$$

This is possible since the focus created by the bent multilayer monochromator is significantly bigger (∼210 μm) than the pinhole collimator itself (75 μm). Moreover, when using other X-ray phase-contrast techniques which are less demanding in terms of spatial coherence (e.g., edge illumination), the same setup can be used with larger collimators potentially delivering a 10 times higher flux.

Of course, despite providing remarkable performances for a such compact design, the integrated photon flux produced by the system presented in this section is more than 3 orders of magnitude smaller than the monochromatic flux achievable at the SYRMEP beamline, in an energy window one order of magnitude broader. In addition, while X-ray spectra produced in synchrotron rings by bending or wiggler magnets are broad, thus allowing a large flexibility in the energy selection, the monochromatic spectra extracted from conventional X-ray sources are limited to the choice of the anode material, therefore to its k-edges.

7.3 Do We Need Clinical Applications in Synchrotrons? A Tentative Answer

Going back to the initial question of this chapter, it is clear that synchrotron radiation facilities offer substantial advantages in X-ray imaging, as demonstrated for the breast CT case in Sect. 7.1, potentially being ideal sources also for clinical applications. On the other hand, a widespread diffusion of SR-based clinical exams is not feasible in terms of costs and infrastructural requirements. For this reason, the diffusion of many phase-contrast techniques, which have the potential to revolutionize X-ray diagnostic, is intrinsically linked to the development of 'synchrotron-like' radiation sources fitting a hospital environment. Therefore, any step forward in the translational research towards more compact sources should be encouraged by all means. In this context, machines based on the inverse Compton scattering [35, 50], which are able of providing sufficiently high coherence and X-ray flux at energies of radiological interest in a scale one or two orders of magnitude smaller than conventional synchrotron facilities, are envisaged as potential candidates to kick off the transition from synchrotrons to hospitals. Anyway, at present, sources of this kind with sufficient robustness and reliability are not available, and high spatial coherence or high output power are mutually exclusive properties of any commercial X-ray device. This dichotomy, ultimately related to the impossibility of dissipating huge heat loads as it would be required for small-focal spot high-power sources, has driven the research down to two separate roads. On one side, sources for medical applications, mainly based on the rotating-anode technology, have been developed pursuing high flux, to speed up the examination, optimized X-ray spectra/detectors, to increase contrast, and sophisticated voltage/current control strategies, to reduce or optimize radiation dose deposition. Typically, these sources have output powers in the order of several kilowatts but they do not feature high brilliance (i.e. number of photon per unit time, area and solid angle) due to the relatively large focal spot size. On the other hand, X-ray imaging laboratory sources, often based on thin transmission or liquid metal anodes, are usually optimized to achieve a small focal spot thus allowing for large geometrical magnifications and/or phase-contrast (mainly propagation-based) imaging. In general, these sources have a small output power in the order of watts but they usually have higher brilliance, the brightest being the ones featuring liquid metal anodes which can tolerate heat loads exceeding the anode's melting point. Right in between these two approaches, several efforts are being dedicated to develop phase-contrast techniques which can be adapted to conventional medical imaging sources. This has been accomplished with some degree of success by using both Talbot-Lau interferometry [51, 52] and edge illumination [53, 54]. Both techniques make use of spatially-varying masks used to split the X-ray beam generated from a broad focal spot into multiple beamlets and to analyze the changes in phase or direction of each beamlet due to the presence of the sample. The presence of absorbing masks brings to a reduction of the X-ray flux, requires for a careful alignment (order of microns) and stability throughout the examination, and demands for a precise fabrication of the masks, which are often made of high-Z materials. The last two conditions are allegedly the most critical issues which, at present, have halted a wider diffusion of these techniques in the clinical context.

In general, synchrotron radiation offers an extremely valuable benchmark and SR-based experiments can provide gold-standards in terms of achievable image quality, defining, in practice, the upper-limit to the potential clinical development of any given technique. At the same time, it is the author's belief that only the successful application of SR-studies on human patients and the production of irrefutable results can trigger the medical community, attracting researchers and funds to make the developed techniques impactful and widely available. Additionally, techniques and technologies born and/or optimized at synchrotrons have not always been confined within large research facilities. As aforementioned, this is the case of phase-contrast techniques as grating interferometry and edge illumination which, firstly implemented at synchrotrons, have been translated to conventional sources. Similarly, propagation-based imaging of human-scale objects could be straightforwardly extended to more compact environments as soon as sources with adequate flux and coherence are available. Finally, it should be noted that the Elettra-based breast CT project described in this work is only one among the several ongoing or planned clinical projects in synchrotron facilities. As mentioned, the researchers of the Australian synchrotron (ANSTO) are developing their own breast CT clinical project [55], planning to start clinical examinations in two years time (2020/2021) and similar interests are also shared by the Indian synchrotron facility (Indus-2) [56]. Along with breast imaging, phase-contrast application to lung imaging has been attracting an increasing interest [57], and encouraging results on human-scale samples have recently appeared in the scientific literature [58]. Historically, besides phase-contrast imaging, one of the most widely investigated medical applications of synchrotron has been the K-edge subtraction technique applied to angiography and/or lung imaging. In this field many clinical systems have been developed over the years at various facilities world-wide as Stanford Synchrotron Radiation Laboratory (SSRL), National Synchrotron Light Source (NSLS), Haburger Synchrotronstrahlungslabor (HASYLAB), Photon Factory (PF), Budker Institute of Nuclear Physics and European Synchrotron Radiation Facility (ESRF) [59]. Moreover, in addition to imaging, clinically-oriented radiotherapy projects [60] are ongoing both at ESRF and ANSTO, while a similar activity is now kicking off at the German Synchrotron (DESY). Therefore, even if the ever-increasing number of synchrotrons is still rather small (around 60 worldwide), an extensive use of these facilities for clinical applications has the potential to provide a relevant clinical impact.

References

1. Brombal L, Arfelli F, Delogu P, Donato S, Mettivier G, Michielsen K, Oliva P, Taibi A, Sechopoulos I, Longo R et al (2019a) Image quality comparison between a phase-contrast synchrotron radiation breast CT and a clinical breast CT: a phantom based study. Sci Rep 9(1):1–12. https://doi.org/10.1038/s41598-019-54131-z
2. Samei E, Krupinski EA (2018) The handbook of medical image perception and techniques. Cambridge University Press. https://doi.org/10.1017/9781108163781
3. O'Connell A, Conover DL, Zhang Y, Seifert P, Logan-Young W, Lin C-FL, Sahler L, Ning R (2010) Cone-beam CT for breast imaging: Radiation dose, breast coverage, and image quality. Am J Roentgenol 195(2):496–509. https://doi.org/10.2214/AJR.08.1017
4. Sechopoulos I, Feng SSJ, D'Orsi CJ (2010) Dosimetric characterization of a dedicated breast computed tomography clinical prototype. Med Phys 37(8):4110–4120. https://doi.org/10.1118/1.3457331
5. Ning R, Conover D, Yu Y, Zhang Y, Cai W, Betancourt-Benitez R, Lu X (2007) A novel cone beam breast CT scanner: system evaluation. In: International society for optics and photonics on medical imaging. Phys Med Imaging 6510:51030. https://doi.org/10.1117/12.710340
6. Benítez RB, Ning R, Conover D, Liu S (2009) NPS characterization and evaluation of a cone beam CT breast imaging system. J X-ray Sci Technol 17(1):17–40. https://doi.org/10.3233/XST-2009-0213
7. Betancourt-Benitez R, Ning R, Conover DL, Liu S (2009) Composite modulation transfer function evaluation of a cone beam computed tomography breast imaging system. Opt Eng 48(11):117002. https://doi.org/10.1117/1.3258348
8. Hernandez AM, Seibert JA, Nosratieh A, Boone JM (2017) Generation and analysis of clinically relevant breast imaging X-ray spectra. Med Phys 44(6):2148–2160. https://doi.org/10.1002/mp.12222
9. Beutel J, Kundel HL, Van Metter RL (2000) Handbook of medical imaging, vol 1. Spie Press. https://doi.org/10.1117/3.832716
10. Gureyev NYI, Timur E (2018) Image quality in attenuation-based and phase-contrast-based X-ray imaging. In: Russo P (ed) Handbook of X-ray imaging: physics and technology. Taylor and Francis, pp 275–305. ISBN 978-1-4987-4152-1. 10.1201/9781351228251
11. Verdun F, Racine D, Ott J, Tapiovaara M, Toroi P, Bochud F, Veldkamp W, Schegerer A, Bouwman R, Giron IH et al (2015) Image quality in CT: from physical measurements to model observers. Phys Med 31(8):823–843. https://doi.org/10.1016/j.ejmp.2015.08.007
12. Solomon JB, Christianson O, Samei E (2012) Quantitative comparison of noise texture across CT scanners from different manufacturers. Med Phys 39(10):6048–6055. https://doi.org/10.1118/1.4752209
13. Dolly S, Chen H-C, Anastasio M, Mutic S, Li H (2016) Practical considerations for noise power spectra estimation for clinical CT scanners. J Appl Clin Med Phys 17(3):392–407. https://doi.org/10.1120/jacmp.v17i3.5841
14. Mizutani R, Saiga R, Takekoshi S, Inomoto C, Nakamura N, Itokawa M, Arai M, Oshima K, Takeuchi A, Uesugi K et al (2016) A method for estimating spatial resolution of real image in the Fourier domain. J Micros 261(1):57–66. https://doi.org/10.1111/jmi.12315
15. Saiga R, Takeuchi A, Uesugi K, Terada Y, Suzuki Y, Mizutani R (2018) Method for estimating modulation transfer function from sample images. Micron 105:64–69. https://doi.org/10.1016/j.micron.2017.11.009
16. Yang K (2018) X-ray cone beam computed tomography. In: Russo P (ed) Handbook of X-ray imaging: physics and technology. Taylor and Francis, pp 713–747. ISBN 978-1-4987-4152-1. 10.1201/9781351228251

17. Bartels M (2013) Cone-beam X-ray phase contrast tomography of biological samples: optimization of contrast, resolution and field of view, vol 13. Universitätsverlag Göttingen. https://doi.org/10.17875/gup2013-92
18. Brombal L, Golosio B, Arfelli F, Bonazza D, Contillo A, Delogu P, Donato S, Mettivier G, Oliva P, Rigon L et al (2018c) Monochromatic breast computed tomography with synchrotron radiation: phase-contrast and phase-retrieved image comparison and full-volume reconstruction. J Med Imaging 6(3):031402. https://doi.org/10.1117/1.JMI.6.3.031402
19. Donato S, Brombal L, Tromba G, Longo R et al (2018) Phase-contrast breast-CT: optimization of experimental parameters and reconstruction algorithms. In: World congress on medical physics and biomedical engineering 2018. Springer, pp 109–115. https://doi.org/10.1007/978-981-10-9035-6_20
20. Brombal L, Donato S, Dreossi D, Arfelli F, Bonazza D, Contillo A, Delogu P, Di Trapani V, Golosio B, Mettivier, G et al (2018b) Phase-contrast breast CT: the effect of propagation distance. Phys Med Biol 63(24):24NT03. https://doi.org/10.1088/1361-6560/aaf2e1
21. Brun F, Brombal L, Di Trapani V, Delogu P, Donato S, Dreossi D, Rigon L, Longo R (2019a) Post-reconstruction 3D single-distance phase retrieval for multi-stage phase-contrast tomography with photon-counting detectors. J Synchrotron Radiat 26(2). https://doi.org/10.1107/S1600577519000237
22. Kalender WA, Kolditz D, Steiding C, Ruth V, Lück F, Rößler A-C, Wenkel E (2017) Technical feasibility proof for high-resolution low-dose photon-counting CT of the breast. Eur Radiol 27(3):1081–1086. https://doi.org/10.1007/s00330-016-4459-3
23. Sarno A, Mettivier G, Russo P (2015) Dedicated breast computed tomography: basic aspects. Med Phys 42(6Part1):2786–2804. https://doi.org/10.1118/1.4919441
24. Kalender WA, Beister M, Boone JM, Kolditz D, Vollmar SV, Weigel MC (2012) High-resolution spiral CT of the breast at very low dose: concept and feasibility considerations. Eur Radiol 22(1):1–8. https://doi.org/10.1007/s00330-011-2169-4
25. Rößler A, Wenkel E, Althoff F, Kalender W (2015) The influence of patient positioning in breast CT on breast tissue coverage and patient comfort. Senologie-Zeitschrift für Mammadiagnostik und-therapie 12(02):96–103. https://doi.org/10.1055/s-0034-1385208
26. Brombal L, Kallon G, Jiang J, Savvidis S, De Coppi P, Urbani L, Forty E, Chambers R, Longo R, Olivo A et al (2019b) Monochromatic propagation-based phase-contrast microscale computed-tomography system with a rotating-anode source. Phys Rev Appl 11(3):034004. https://doi.org/10.1103/PhysRevApplied.11.034004
27. Wilkins S, Nesterets YI, Gureyev T, Mayo S, Pogany A, Stevenson A (2014) On the evolution and relative merits of hard X-ray phase-contrast imaging methods. Phil Trans R Soc A 372(2010):20130021. https://doi.org/10.1098/rsta.2013.0021
28. Olivo A, Castelli E (2014) X-ray phase contrast imaging: from synchrotrons to conventional sources. Rivista del nuovo cimento 37(9):467–508. https://doi.org/10.1393/ncr/i2014-10104-8
29. Rigon L (2014) X-ray imaging with coherent sources. In Brahme A (ed) Comprehensive biomedical physics, vol 2. Elsevier, 193–216. https://doi.org/10.1016/B978-0-444-53632-7.00209-4
30. Bravin A, Coan P, Suortti P (2012) X-ray phase-contrast imaging: from pre-clinical applications towards clinics. Phys Med Biol 58(1):R1. https://doi.org/10.1088/0031-9155/58/1/R1
31. Cosslett V, Nixon W (1951) X-ray shadow microscope. Nature 168(4262):24–25. https://doi.org/10.1038/168024a0
32. Mayo S, Davis T, Gureyev T, Miller P, Paganin D, Pogany A, Stevenson A, Wilkins S (2003) X-ray phase-contrast microscopy and microtomography. Opt Express 11(19):2289–2302. https://doi.org/10.3109/02841851.2010.504742
33. Fella C, Balles A, Zabler S, Hanke R, Tjeung R, Nguyen S, Pelliccia D (2015) Laboratory X-ray microscopy on high brilliance sources equipped with waveguides. J Appl Phys 118(3):034904. https://doi.org/10.1063/1.4927038
34. Sowa KM, Jany BR, Korecki P (2018) Multipoint-projection X-ray microscopy. Optica 5(5):577–582. https://doi.org/10.1364/OPTICA.5.000577

35. Gradl R, Dierolf M, Hehn L, Günther B, Yildirim AÖ, Gleich B, Achterhold K, Pfeiffer F, Morgan KS (2017) Propagation-based phase-contrast X-ray imaging at a compact light source. Sci Rep 7(1):4908. https://doi.org/10.1038/s41598-017-04739-w

36. Töpperwien M, Gradl R, Keppeler D, Vassholz M, Meyer A, Hessler R, Achterhold K, Gleich B, Dierolf M, Pfeiffer F et al (2018) Propagation-based phase-contrast X-ray tomography of cochlea using a compaCT synchrotron source. Sci Rep 8(1):4922. https://doi.org/10.1038/s41598-018-23144-5

37. Tuohimaa T, Otendal M, Hertz HM (2007) Phase-contrast X-ray imaging with a liquid-metal-jet-anode microfocus source. Appl Phys Lett 91(7):074104. https://doi.org/10.1063/1.2769760

38. Krenkel M, Töpperwien M, Dullin C, Alves F, Salditt T (2016) Propagation-based phase-contrast tomography for high-resolution lung imaging with laboratory sources. AIP Adv 6(3):035007. https://doi.org/10.1063/1.4943898

39. Vittoria FA, Endrizzi M, Kallon GK, Hagen CK, Iacoviello F, De Coppi P, Olivo A (2017) Multimodal phase-based X-ray microtomography with nonmicrofocal laboratory sources. Phys Rev Appl 8(6):064009. https://doi.org/10.1103/PhysRevApplied.8.064009

40. Kallon G, Diemoz P, Vittoria F, Basta D, Endrizzi M, Olivo A (2017) Comparing signal intensity and refraction sensitivity of double and single mask edge illumination lab-based X-ray phase contrast imaging set-ups. J Phys D Appl Phys 50(41):415401. https://doi.org/10.1088/1361-6463/aa8692

41. Shimizu K, Omote K (2008) Multilayer optics for X-ray analysis. Rigaku J 24(1)

42. Oberta P, Platonov Y, Flechsig U (2012) Investigation of multilayer X-ray optics for the 6 kev to 20 kev energy range. J Synchrotron Radiat 19(5):675–681. https://doi.org/10.1107/S0909049512032153

43. Samei E, Flynn MJ, Reimann DA (1998) A method for measuring the presampled MTF of digital radiographic systems using an edge test device. Med Phys 25(1):102–113. https://doi.org/10.1118/1.598165

44. Van Nieuwenhove V, De Beenhouwer J, De Carlo F, Mancini L, Marone F, Sijbers J (2015) Dynamic intensity normalization using eigen flat fields in X-ray imaging. Opt Express 23(21):27975–27989. https://doi.org/10.1364/OE.23.027975

45. Brun F, Pacilè S, Accardo A, Kourousias G, Dreossi D, Mancini L, Tromba G, Pugliese R (2015) Enhanced and flexible software tools for X-ray computed tomography at the italian synchrotron radiation facility elettra. Fundamenta Informaticae 141(2–3):233–243. https://doi.org/10.3233/FI-2015-1273

46. Henke BL (2018) CXRO X-ray interaction with matter. http://henke.lbl.gov/optical_constants

47. Totonelli G, Maghsoudlou P, Georgiades F, Garriboli M, Koshy K, Turmaine M, Ashworth M, Sebire NJ, Pierro A, Eaton S et al (2013) Detergent enzymatic treatment for the development of a natural acellular matrix for oesophageal regeneration. Pediatr Surg Int 29(1):87–95. https://doi.org/10.1007/s00383-012-3194-3

48. Hagen CK, Maghsoudlou P, Totonelli G, Diemoz PC, Endrizzi M, Rigon L, Menk R-H, Arfelli F, Dreossi D, Brun E et al (2015) High contrast microstructural visualization of natural acellular matrices by means of phase-based X-ray tomography. Sci Rep 5:18156. https://doi.org/10.1038/srep18156

49. Scotton CJ, Hayes B, Alexander R, Datta A, Forty EJ, Mercer PF, Blanchard A, Chambers RC (2013) Ex vivo μCT analysis of bleomycin-induced lung fibrosis for pre-clinical drug evaluation. Eur Respir J erj01824–2012. https://doi.org/10.1183/09031936.00182412

50. Eggl E, Dierolf M, Achterhold K, Jud C, Günther B, Braig E, Gleich B, Pfeiffer F (2016) The munich compaCT light source: initial performance measures. J Synchrotron Radiat 23(5):1137–1142. https://doi.org/10.1107/S160057751600967X

51. Pfeiffer F, Weitkamp T, Bunk O, David C (2006) Phase retrieval and differential phase-contrast imaging with low-brilliance X-ray sources. Nat Phys 2(4):258. https://doi.org/10.1038/nphys265

52. Arboleda C, Wang Z, Jefimovs K, Koehler T, Van Stevendaal U, Kuhn N, David B, Prevrhal S, Lång K, Forte S et al (2019) Towards clinical grating-interferometry mammography. Eur Radiol 1–7. https://doi.org/10.1007/s00330-019-06362-x

53. Endrizzi M, Diemoz PC, Millard TP, Louise Jones J, Speller RD, Robinson IK, Olivo A (2014) Hard X-ray dark-field imaging with incoherent sample illumination. Appl Phys Lett 104(2):024106. https://doi.org/10.1063/1.4861855
54. Havariyoun G, Vittoria FA, Hagen CK, Basta D, Kallon GK, Endrizzi M, Massimi L, Munro PR, Hawker S, Smit B et al (2019) A compaCT system for intraoperative specimen imaging based on edge illumination X-ray phase contrast. Phys Med Biol. https://doi.org/10.1088/1361-6560/ab4912
55. Gureyev T, Nesterets YI, Baran P, Taba S, Mayo S, Thompson D, Arhatari B, Mihocic A, Abbey B, Lockie D et al (2019) Propagation-based X-ray phase-contrast tomography of mastectomy samples using synchrotron radiation. Med Phys. https://doi.org/10.1002/mp.13842
56. Sharma R, Sharma S, Sarkar P, Singh B, Agrawal A, Datta D (2019) Phantom-based feasibility studies on phase-contrast mammography at indian synchrotron facility indus-2. J Med Phys 44(1):39. https://doi.org/10.4103/jmp.JMP_98_18
57. Kitchen MJ, Buckley GA, Gureyev TE, Wallace MJ, Andres-Thio N, Uesugi K, Yagi N, Hooper SB (2017) CT dose reduction factors in the thousands using X-ray phase contrast. Sci Rep 7(1):15953. https://doi.org/10.1038/s41598-017-16264-x
58. Wagner WL, Wuennemann F, Pacilé S, Albers J, Arfelli F, Dreossi D, Biederer J, Konietzke P, Stiller W, Wielpütz MO et al (2018) Towards synchrotron phase-contrast lung imaging in patients—a proof-of-concept study on porcine lungs in a human-scale chest phantom. J Synchrotron Radiat 25(6). https://doi.org/10.1107/S1600577518013401
59. Thomlinson W, Elleaume H, Porra L, Suortti P (2018) K-edge subtraction synchrotron X-ray imaging in bio-medical research. Phys Med 49:58–76. https://doi.org/10.1016/j.ejmp.2018.04.389
60. Grotzer M, Schültke E, Bräuer-Krisch E, Laissue J (2015) Microbeam radiation therapy: clinical perspectives. Phys Med 31(6):564–567. https://doi.org/10.1016/j.ejmp.2015.02.011

Chapter 8
Conclusions

The work substantiating this thesis has contributed to add some of the missing pieces towards the clinical implementation of the propagation-based phase-contrast breast CT at Elettra, in the framework of the SYRMA-3D collaboration. The project has the ambitious goal of integrating a not yet widespread radiological technique such as breast CT into a synchrotron facility environment, proving, in a specific context, the advantages of phase-contrast imaging and its diagnostic impact on one of the most challenging imaging tasks: early breast cancer detection.

As the realization of the project requires to address several multifaceted problems, the range of topics and issues covered in this work has been quite broad, spanning from detector performance to fundamental physical modeling of image quality metrics. Specifically, the presence of detector-related artifacts in tomographic reconstructions has been tackled via a dedicated pre-processing procedure containing suitable interpolation techniques to compensate for insensitive gaps between adjacent detector modules and time-dependent gain variations due to charge-trapping effects (Chap. 4). The need for optimization of the experimental setup has led to an in-depth study of signal and noise propagation through the whole imaging chain, allowing for the first time to achieve an accurate matching, in terms of signal-to-noise ratio gain due to phase retrieval, between theoretical predictions and experimental images as a function of propagation distance and pixel size (Chap. 5, Sects. 5.1, 5.2). The outcomes of this analysis have led to the design of an extension of the SYRMEP beamline which, when installed, will allow to obtain images with increased signal-to-noise ratio (by a factor of 2 or more) at the present radiation dose level. In the same context, pursuing the goal of reducing the scan time for large volumes while delivering a more uniform dose distribution, a new filtration system has been developed to use a wider portion of the incoming Gaussian X-ray beam (50% wider), while uniforming its spatial intensity distribution (Chap. 5, Sect. 5.3). Additional effort has been put into data-processing, implementing a post-reconstruction phase-retrieval

L. Brombal, *X-Ray Phase-Contrast Tomography*, Springer Theses,
https://doi.org/10.1007/978-3-030-60433-2_8

procedure allowing to compensate for periodic artifacts in the reconstructed volume, in case of acquisitions requiring multiple vertical translations (Chap. 5, Sect. 5.4). The aforementioned results, despite being mostly finalized to the breast CT implementation, have a rather general applicability to many synchrotron radiation and/or propagation-based imaging setups.

Several large surgical breast specimens have been scanned at clinically compatible dose levels and the resulting images have been compared with clinical mammography, showing, for instance, increased sensitivity in microcalcification detection and a better depiction of lesions morphology (Chap. 6). To directly assess and demonstrate the advantages of propagation-based breast CT over conventional systems, the performances of the developed setup have been tested against one commercially available and clinically used breast CT system, thanks to the collaboration with the Radboud University Medical Center (Nijmegen, The Netherlands). The results of this first-of-its-kind quantitative comparison study (Chap. 7, Sect. 7.1) indicate that synchrotron-based imaging yields major advantages in terms of signal-to-noise ratio (higher by a factor up to 3), spatial resolution (higher by a factor up to 5) and detail visibility, thus providing a further justification for the realization of the SYRMA-3D project. In addition, thanks to the collaboration with the Department of Medical Physics and Biomedical Engineering, University College London (London, UK), the scientific horizon of the thesis has been widened to a laboratory implementation of propagation-based micro-CT based on a high-power rotating anode source (Chap. 7, Sect. 7.2); results show that the phase-contrast signal can be higher than attenuation contrast (up to a factor of 3), and quantitative (monochromatic) CT images of samples of bio-medical interest (i.e. esophageal tissue and lung tissue) can be obtained in scan times ranging from some minutes to few hours, demonstrating that rotating anode sources can be valuable and reliable tools also for propagation-based imaging laboratory applications.

Many results presented throughout this work have already been documented in 9 separate publications on scientific journals covering a wide spectrum of topics spanning from medical physics (Physics in Medicine and Biology), to applied physics (Physical Review Applied, Scientific Reports), synchrotron physics (Journal of Synchrotron Radiation) and scientific instrumentation (Journal of Instrumentation).

Even if it is clear that radiological applications in synchrotrons, as the one presented in this work, cannot reach a wide population, this kind of studies offer valuable benchmarks and directly prove the diagnostic benefits of phase-contrast imaging. This is of great importance especially when interfacing with the medical community, which often gives more credit to few but clinically relevant results rather than many theoretical or proof-of-principle speculations. Within this framework, SYRMA-3D is only one among the several ongoing or planned clinical projects in synchrotron facilities. Anyway, it is the author's belief that an even wider diffusion of such applications is key for reaching the critical mass of experienced scientists and medical doctors which is needed to trigger the long-anticipated transition of phase contrast from synchrotrons to hospitals, ultimately bringing to a better X-ray diagnostic available to a large number of people.

Appendix A
Equivalence of TIE and Ray-Tracing Approaches

In this appendix the equivalence between the X-ray intensity reaching the detector plane computed through a ray tracing approach, Eq. (2.12), and the transport-of-intensity equation (TIE), Eq. (2.16), is demonstrated.

The TIE reads

$$\nabla_{xy}\left[I(x, y; z = 0)\nabla_{xy}\Phi(x, y; z = 0)\right] = -k\frac{\partial I(x, y; z = 0)}{\partial z} \tag{A.1}$$

where I expresses the X-ray intensity as a function of the position x, y at the object plane $z = 0$, Φ is the phase shift, k the wave number and ∇_{xy} the gradient operator in the transverse plane. By further performing the finite-difference approximation

$$\frac{\partial I(x, y; z = 0)}{\partial z} \simeq \frac{I(x, y; z = z_1) - I(x, y; z = 0)}{z_1} \tag{A.2}$$

where z_1 is the image plane coordinate (i.e. propagation distance), TIE can be re-written as

$$I(x, y; z = z_1) = I(x, y; z = 0) - \frac{z_1}{k}\nabla_{xy}\left[I(x, y; z = 0)\nabla_{xy}\Phi(x, y; z = 0)\right] \tag{A.3}$$

The last term of the previous equation can be approximated as

$$\nabla_{xy}\left[I(x, y; z = 0)\nabla_{xy}\Phi(x, y; z = 0)\right] = I(x, y; z = 0)\nabla_{xy}^2\Phi(x, y; z = 0)$$
$$+ \nabla_{xy}I(x, y; z = 0)\nabla_{xy}\Phi(x, y; z = 0)$$
$$\simeq I(x, y; z = 0)\nabla_{xy}^2\Phi(x, y; z = 0) \tag{A.4}$$

which is valid if the transverse phase gradient and/or the transverse intensity gradient is not too strong. Specifically, the latter condition is reasonable when imaging soft

L. Brombal, *X-Ray Phase-Contrast Tomography*, Springer Theses, https://doi.org/10.1007/978-3-030-60433-2

tissues since, in the object plane, strong intensity variations are not present due to the poor attenuation contrast of such samples. At this point Eq. (A.3) can be written as

$$I(x, y; z = z_1) = I(x, y; z = 0)\left[1 - \frac{z_1}{k}\nabla_{xy}^2 \Phi(x, y; z = 0)\right] \qquad (A.5)$$

which, identifying $I(x, y; z = 0) = I_0 e^{-2k \int \beta(x,y,z)\,dz}$ (β is the imaginary part of the refractive index) as the X-ray beam intensity emerging from the sample, is identical to Eq. (2.12), *Q.E.D.*

Appendix B
Normalization Factor in Eq. 4.11

The presence of the normalization factor M/M_0 in Eq. (5.11) is due to the fact that the experimental data were collected keeping constant the fluence at the detector plane instead of the sample plane. As reported in Eq. (5.5), when no PhR is applied, the variance dependence on the effective pixel size $h' = h/M$ and the X-ray fluence at the object Φ is

$$\text{var} \propto \frac{1}{h'^4 \Phi} = \frac{1}{\frac{h^4}{M^4} \Phi} \propto \frac{M^4}{\Phi} \tag{B.1}$$

where $h = 60$ um is the physical pitch, which is fixed, and M is the geometrical magnification. Considering an X-ray source emitting a given number of photons per unit solid angle ϕ, the fluence is written as

$$\Phi = \frac{\phi}{z_0^2} \tag{B.2}$$

where z_0 is the source-to-sample distance and ϕ is a constant property of the source. By recalling the definition of geometrical magnification, z_0 can be written as

$$z_0 = \frac{z_0 + z_1}{M} \propto \frac{1}{M} \tag{B.3}$$

where $z_0 + z_1$ gives the source-to-detector distance which, in the experimental setup described in Sect. 5.1.2, is a constant. The latter equation implies that

$$\Phi \propto M^2 \tag{B.4}$$

and, by inserting this result in Eq. (B.1)

$$\text{var} \propto \frac{M^4}{M^2} = M^2 \tag{B.5}$$

© The Editor(s) (if applicable) and The Author(s), under exclusive license
to Springer Nature Switzerland AG 2020
L. Brombal, *X-Ray Phase-Contrast Tomography*, Springer Theses,
https://doi.org/10.1007/978-3-030-60433-2

At this point, if measured without any normalization, the signal-to-noise ratio, which is inversely proportional to image noise, would be

$$\text{SNR}_{\text{noNorm}} \propto \frac{1}{\sqrt{\text{var}}} \propto \frac{1}{M} \tag{B.6}$$

For this reason, with the aim of highlighting the sole effect of phase-retrieval eliminating the contribution of geometrical magnification, the SNR in Eq. (5.11) contains the normalization factor

$$\text{SNR} = \text{SNR}_{\text{noNorm}} \frac{M}{M_0} \tag{B.7}$$

where M_0 is a constant and small (1.05) magnification corresponding to the patient support position which is used as a reference.

Appendix C
Derivation of Eq. 6.7

In this appendix the relationship between the full-width-at-half-maximum (FWHM) of a Gaussian point spread function (PSF) and the frequency at 10% of the corresponding modulation transfer function (MTF) is demonstrated.

A Gaussian PSF is written as

$$\text{PSF}(x) = \frac{1}{\sigma\sqrt{2\pi}} \exp\left(-\frac{x^2}{2\sigma^2}\right) \tag{C.1}$$

where its FWHM is proportional to the standard deviation σ through the formula

$$\text{FWHM} = 2\sqrt{2\log(2)}\sigma \tag{C.2}$$

The corresponding MTF, function of the spatial frequency f will be:

$$\text{MTF}(f) = |\mathscr{F}\left[\text{PSF}\right](f)| = \exp\left(-\frac{(2\pi f)^2\sigma^2}{2}\right) = \exp\left(-\frac{(2\pi f)^2\text{FWHM}^2}{2\left(2\sqrt{2\log(2)}\right)^2}\right) \tag{C.3}$$

where $\mathscr{F}$ denotes the Fourier transform. To find the frequency corresponding to the 10% amplitude of the MTF, $f_{10\%}$, means to invert the equation

$$10\% = \exp\left(-\frac{(2\pi f_{10\%})^2\text{FWHM}^2}{2\left(2\sqrt{2\log(2)}\right)^2}\right) \tag{C.4}$$

which, as reported in Eq. (7.7), results in

$$f_{10\%} = \frac{2}{\pi}\sqrt{\log(10)\log(2)}\frac{1}{\text{FWHM}} \simeq \frac{1}{1.24 \times \text{FWHM}} \ . \tag{C.5}$$

© The Editor(s) (if applicable) and The Author(s), under exclusive license to Springer Nature Switzerland AG 2020
L. Brombal, *X-Ray Phase-Contrast Tomography*, Springer Theses,
https://doi.org/10.1007/978-3-030-60433-2

Author's Curriculum Vitae

Luca Brombal

Dept. of Physics, University of Trieste 🏢
via A. Valerio 2, 34127, Trieste, Italy 📍
lbrombal@units.it ✉
researchgate.net/profile/Luca_Brombal 🌐
born 04/06/1992 📅

Research

2020–date   **Post-Doctoral Researcher**
University of Trieste

- **K-edge subtraction imaging**: working at the implementation of state-of-art bent-Laue crystal monochromator within the INFN-funded K-edge Imaging at Synchrotron Sources (KISS) project.

- **Phase-contrast breast CT**: finalizing the optimization/ implementation of the first propagation-based phase-contrast breast CT with synchrotron radiation. Focus on quantitative models of image quality vs propagation distance and motion-related artifacts compensation.

2019   **Researcher**
              Elettra Sincrotrone S.C.p.A

- **Analyzer-based imaging**: improving the image-reconstruction process via an in-depth characterization of the system. Implemented simulation of X-ray scattering from spheres through multiple refraction process in Geant4 toolkit.

- **Users' support**: SYRMEP beamline as scientific support to users during micro-tomography and phase-contrast imaging beamtimes.

2018   **Visiting Scientist (Erasmus+ Traineeship)**
              University College London (UCL)

- **Edge-illumination imaging**: first of its kind development of single-mask edge illumination by using a photon-counting detector and high-power conventional X-ray source.

- **Phase-contrast CT**: SYRMEP beamline as scientific support to users during micro-tomography and phase-contrast imaging beamtimes.

2016–2019   **Ph.D. student**
              University of Trieste

- **Phase-contrast Breast CT**: working on the INFN-funded SYRMA-3D project in the development of a phase-contrast breast CT synchrotron-based facility. Focus on CT data acquisition, image quality analysis, modelling of the imaging chain, photon-counting detector-specific pre-processing, hardware and software development.

Education

2016–2020   **Ph.D. in Physics (Doctor Europaeus cum laude)**
              University of Trieste
              Thesis title: *X-ray Phase-Contrast Tomography: Underlying Physics and Developments for Breast Imaging*

2014–2016   **M.Sc in Physics (honours—110/110 e lode)**
University of Trieste
Thesis title: *PET data for the monitoring of protontherapy treatments*

2011–2014   **B.Sc in Physics (honours—110/110 e lode)**
University of Trieste
Thesis title: *Spatial localization techniques in MRI spectroscopy for clinical applications*

Affiliations and Collaborations

2016–date   **National Institute of Nuclear Physics (INFN) associate**
INFN - Division of Trieste

2017–date   **Member of the SYRMEP beamline research group**
Elettra Sincrotrone S.C.p.A.
SYRMEP Group

2018–date   **Collaborator of the Advanced X-ray Imaging group**
Dept. of Medical Physics and Biomedical Engineering - UCL
AXiM group

2019–date   **Collaborator of EMITEL (e-Encyclopedia of Medical Physics and Multilingual Dictionary of Terms)**
Author of the first 9 items related to X-ray phase-contrast imaging.
emitel2.eu

Selected Publications

Brombal L (2020) Effectiveness of X-ray phase-contrast tomography: effects of pixel size and magnification on image noise. J Instrum 15(01):C01005.

Brombal L et al (2019) Image quality comparison between a phase-contrast synchrotron radiation breast CT and a clinical breast CT: a phantom based study. Scient Reports 9(1):1–12.

Brombal L et al (2019) Monochromatic propagation-based phase-contrast microscale computed-tomography system with a rotating-anode source. Phys Rev Appl 11(3):034004.

Brombal L et al (2018) Phase-contrast breast CT: the effect of propagation distance. Phys Medic Biol 63(24):24NT03.

Brombal L et al (2018) Large-area single-photon-counting CdTe detector for synchrotron radiation computed tomography: a dedicated pre-processing procedure. J Synchrotron Radiat 25(4).

Brombal L et al (2018) Monochromatic breast CT: absorption and phase-retrieved images. Medical Imaging 2018: Physics of Medical Imaging, vol. 10573. International Society for Optics and Photonics.

Brombal L et al (2017) Proton therapy treatment monitoring with in-beam PET: investigating space and time activity distributions. Nuclear instruments and methods in physics research section a: accelerators, spectrometers, detectors and associated equipment 861:71–76.

Full publications list at arts.units.it

Printed in the United States
by Baker & Taylor Publisher Services